THE LITTLE Herb ENCYCLOPEDIA REVISED

By: Jack Ritchason, N.D.

BiWorld Publishers
P.O. Box 1143
Orem, Utah 84057

About the Author

Dr. John (Jack) Ritchason has been in the health field for 19 years and has lectured nationally and internationally on herbs, vitamins, nutrition, and iridology. He graduated as a Naturopathic Doctor from Arizona College of Naturopathic Medicine, a branch of American University of natural Therapeutics and Preventive Medicine. He is also a Registered Healthologist, Iridologist, Touch-for-Health Instructor, and Herbalist. He is a Senior National Manager for Nature's Sunshine and has his Ph.D from Donsbach University.

DEDICATION

This book is dedicated to my very dear and loving wife who has always inspired me to continue seeking knowledge in all fields, especially health. Without her this book would not be possible. Her endless hours of typing and correcting my bad grammar and spelling were badly needed. It is my hope that this book will become a guideline to my posterity and others in their search for a better and healthier life. I am thankful that I have an all-just and merciful Heavenly Father who has given us the answer already for our physical problems on this earth. We simply need to search for the knowledge that is already here, for without health, we have nothing.

Jack Ritchason, N.D., Ph.D.

ISBN 0-89557-059-9
BiWorld Publishers
P.O. Box 1143
Orem, Utah 84057

CONTENTS

INTRODUCTION

Life was meant to be full and fruitful. God endowed man with the ability to feel, to smell, to taste. He intended that we experience, savor, and enjoy the multitude of his creations. But first we must enjoy the most fundamental of all blessings—good health. We must be whole.

This book is designed to aid you in your quest for good health. Just as God intended that we find it, he also gave us a compass. All we need do is but follow in the right direction. "And God said, Behold, I have given you every herb bearing seed, which is upon the face of all the earth, and every tree, in the which is the fruit of a tree yielding seed; to you it shall be for meat." (Genesis 1:29)

God considers our bodies to be holy temples, and their misuse or neglect will be to our condemnation. "If any man defile the temple of God, him shall God destroy; for the temple of God is holy, which temple ye are." (1 Corinthians 3:17.) Ignorance can no longer be an excuse. The responsibility is clear, and now, at last, so also is the path to achieve what we desire most, good health. Health is wealth and without health we have nothing.

How to Prepare and Use Herbs

One of the wonders of herbs is their tremendous versatility, not only for the scores of maladies which they treat, but for the number of ways in which they can be prepared and used. They are most commonly associated with teas. A decoction is a tea made from roots and bark. To prepare a decoction, gently simmer approximately 1 teaspoon of powdered herb or 1 tablespoon of cut herb in 1 cup of water for 30 minutes. Allow the tea to stand for 10 minutes before drinking. An infusion is also a tea; however, it is made from leaves and blossoms. It is prepared by steeping 1 teaspoon of powdered herb in 1 cup of boiling water for 10 minutes.

Oil of herbs is an extraction of an herb in an oil base. This is prepared by placing the powdered herb in a double boiler and covering it with olive oil. The mixture should be simmered for approximately 3 hours, then strained. A tincture is similar except that it is an extraction in vinegar or alcohol. Tinctures should be prepared as directed for the specific herb.

For external use, a poultice or herb pack is most helpful. A poultice is prepared by mixing crushed fresh herbs or powdered herbs with mineral water to form a thick paste. The mixture is spread on a clean cloth and placed over the affected area.

Medical science has been a great boon to herbal medicine in that many herbs can now be purchased in pure gelatin capsules—undoubtedly the easiest method for taking large herb doses. When taking herb capsules and tablets, most herbalists agree that you should always take them with a full glass of liquid and before eating. It is recommended that when a person takes herbs in a tablet form that they chew them thoroughly so it is guaranteed that they will be broken up in the stomach to aid in digestion. Many people taking compressed tablets pass them whole.

Herbal Medicine

Regardless of their preparation, when used judiciously and properly, herbs bring new health to the lives of many people. Their value as a medicine has been known for centuries. Herbal remedies were the natural drugs used to cure various ailments of long ago. King Solomon used herbs. In fact, common references to herbs throughout the Bible indicate that herbs were used by kings and commoners alike.

Today's health-conscious public is now realizing that herbs can also bring all-around better living. Athletes are finding that certain herbs give added strength and stamina. Herbs are helping thousands to control weight and keep in shape. As a beauty aid they are far surpassing any of the concoctions of today's cosmetologists. Students are finding that herbs enhance their alertness and mental capacity. But is it any wonder? "He causeth the grass to grow for the cattle, and herb for the service of man; that he may bring forth food out of the earth." (Psalm 104:14)

Good health comes from good food, from adequate exercise, and from giving special attention to particular areas of concern. With health as our goal and proper nutrition as our tool, we will find the peace without and within ourselves that the world seeks. *The Little Herb Encyclopedia Revised* is here to help achieve that peace, that balance, that whole life.

HERB REMEDIES

Natural health care is coming of age—again. *The Little Herb Encyclopedia Revised* lists the most traditional, accepted, and widely successful herbal remedies for common health problems.

Many different herb remedies have been successfully used to cure health problems. The remedies explained in this book are the most common and most successful remedies. But they are not the only remedies. The book is a "first step" guide that tells how most people have overcome their poor health conditions. We offer only advice that has worked for others. it is not our intent to diagnose or prescribe.

And in the event of a medical emergency we definitely recommend seeking medical help. Herbs are a slow, safe cure. More drastic steps are usually needed in an emergency. In case of an emergency, unless you are a Master Herbalist, you should seek competent help.

CLEAN AND DETOXIFY

To understand how to use herbs to restore health, we must first understand how foods affect the body. Proteins, vitamins and minerals are the builders and maintainers of the cells. Herbs are the cleansers of the body, feeders of the glands and balancers of the hormones. Unfortunately, no matter how well we care for our bodies, they naturally build up toxins. Toxins are poisonous substances which are the result of metabolic activity. Those toxins which the body cannot eliminate are often detoxified by the building of antibodies. If not, illness ensues.

Herbs make the work of our bodies simpler by removing the toxins. For this reason it is understandable that when an herb is used for the first time, it may have a laxative effect. This is a natural reaction as the body cleanses itself of impurities. If the laxative effect is too great, reduce the herb intake.

A disease crisis happens to save your life when the body is full of toxins and excess mucus. A disease crisis is a cold, flu, and so on, and is usually accompanied by a fever. It lasts one to two weeks or longer. There is no need to do anything but quit eating, except for water or potassium broth, and if the bowels are not moving properly, take an enema. The person is usually constipated.

A healing crisis occurs when the body is strong enough to stand the cleansing process and when you feel your best. It usually lasts one to three days or one to two weeks at most. This happens after eating correctly for about a three month period. Potassium is also beneficial and generally needed.

HERBS, THE SAFE REMEDY

When using a new herb, begin with a recommended dosage and observe how your body reacts to it. If you do not receive the desired result, increase or decrease the dosage accordingly. When herbs are used for restoring health, dosages will need to be five-to-seven times that which is required for maintaining health. Depending of course on the ailment and the severity of it, approximately three months of sustained high dosage is usually required before the body is repaired. Sometimes it takes a full year for reparation to occur so that a person goes through all four seasons. The science of Homeopathy teaches that it takes one month for every year of illness for the person to rebuild the body.

Medical science has made us naturally wary of overdoses and wrong dosages. With herbs you can put your mind at ease. Herbs are food, not drugs. There are no dangerous side effects as with drugs. It takes time and experimenting to learn which herbs work best for you. Years of misuse or neglect cannot be repaired any more readily with herbs than with dangerous drugs. Give herbs a chance. Allow the body to heal slowly, naturally, and effectively with herbs.

Proper food is very important and some people think that all they have to do is take a few herbs and everything will be all right. But, if they will not correct their choice of food that they are consuming every day, their problems will not stay away.

HERBS DURING PREGNANCY

Most herbs can be taken throughout pregnancy with no ill effects. Many herbs are helpful during pregnancy, especially just before childbirth. The herbs that are helpful can be found under the appropriate headings in the remedies. Helpful herbs would include those to cure nausea, morning sickness, pain, and the like.

Many herbalists strongly recommend that some herbs NOT be taken during pregnancy. These herbs are black cohosh, pennyroyal and dong quai. Most of the hormone herbs mentioned above should NOT be taken.

HERBS FOR CHILDREN

Children can take most herbs with no adverse effects. Obviously, the dosages for medicinal herbs must be reduced since the child is smaller than an adult. A general rule for determining dosage is to calculate the dosage according to the body weight of the child. Preschool children take approximately one forth an adult dosage. From age five through ten, one half the adult dosage is sufficient. Early teens take three-fourths the adult dosage. When the growing child reaches adult size (sometime between the ages of 13 and 20), then full doses can be used.

Many herbalists urge that children take nutrient herbs throughout the growing years to help build strong bodies. These nutrient herbs can usually be taken in relatively large amounts with no ill effects whatsoever. The recommended food herbs include alfalfa, rose hips, dandelion, comfrey, bee pollen, capsicum, garlic, horsetail, kelp, papaya, red raspberry, slippery elm, and thyme. Scullcap, gotu kola, and eyebright are also popular as diet supplements to aid learning and studying during school age years. Many herbalists recommend that children take a calcium rich herb combination (see page 81) for growing bodies and bones.

It is often difficult to give needed medicinal herbs to infants and small children because the children are unable to swallow the capsules. The herb powder is sometimes so bitter that it cannot be added to milk or food; some mothers have used bananas, applesauce, or peanut butter to hide the bitter taste of the herbs. When the taste is so bitter that it cannot be hidden, the herb can be taken through enemas or as suppositories in the rectum.

Although some of the aromatic herbs are very delicious and many are high in nutritive value, the only herbs to be cautious about giving to children are the hormone herbs. Ginseng, damiana, and black cohosh should not be given to children before they reach puberty. Also, some of the strong laxative herbs and herb combinations should be used sparingly. Cascara sagrada is one herb that can have strong laxative effects for children.

NOT PRESCRIPTIONS

The Little Herb Encyclopedia Revised does not directly or indirectly dispense medical advice or prescribe the use of herbs as a form of treatment for sickness without medical approval. Nutritionists and other experts in the field of health hold widely varying views. It is not the intent of this book to diagnose or prescribe. The information is a collection of various treatments and uses of herbs and herb combinations from a variety of sources. Only the most common or most widely accepted uses are mentioned. If you decide to diagnose for yourself and use the information without your doctor's approval, you will be prescribing for yourself, which is your constitutional right, but the author assumes no responsibility.

ACNE

Popular treatments include purifying the blood with an herbal blood purifier combination, eating fresh fruits, keeping the bowels moving with lower bowel cleanser herb combination. Refrain from eating too many sweets and greasy foods. Use herbs high in vitamin A. (See *Little Vitamin and Mineral Encyclopedia* for this information.)

AFTERPAIN

An herb combination for pain is recommended. A calcium rich herb intake relieves the pain for people also. Red raspberry is recommended through pregnancy.

AGE SPOTS

Believed to be caused by liver malfunction, so a liver corrective herb formula is recommended along with vitamins E, C, A, and licorice.

ALLERGIES

Bee pollen, comfrey, and an asthma corrective herb

combination are often used. Vitamins A, C, and E help. Sinus problems are resolved by using sinus herb combination. It is generally beneficial to clease the lymph system and the bowel.

ANEMIA

Chlorophyll, yellow dock, and dandelion are often recommended to clean the blood, build nutritive salts, and replenish iron. Also an iron-rich formula containing red beet, yellow dock, strawberry leaves, lobelia, burdock root, nettle, mullein leaves has been used. Vitamin E is also used.

APPETITE

Saw palmetto, dandelion, and barberry are often used to increase the appetite. Fennel, peppermint, and camomile depress the appetite and calm the stomach.

ARTERIOSCLEROSIS

A capsicum and garlic combination is often used with lecithin and a heart herb combination as well. Replace salt with kelp. There are also very effective herb chelation formulas which are blood purifiers as well.

ARTHRITIS

An arthritis herb combination was created especially for treatment of this problem and is often used with a special herbal formula. Usually it is believed that calcium should be replenished with calcium rich digestate (hydrochloric acid) to help digest some foods and especially calcium, which is needed for repair. Using an herbal blood purifier combination also is believed to be helpful.

ASTHMA

An asthma corrective herb combination is very useful for this. Other often recommended herbs include combination comfrey and fenugreek, and the single herb lobelia. Cleansing the bowel is also helpful.

BACK PAIN

A thorough cleansing of the lower intestinal area by using a lower bowel cleanser herb combination helps many people. A garlic and catnip enema is another popular remedy. For the pain, an herbal combination for pain, an herb combination formula for nerves, and a safe sleep herb formula are utilized. Building up calcium levels with a calcium rich herb combination helps many sufferers, as 88 percent of lower back pain is caused by calcium deficiencies.

BAD BREATH

Believed to be caused by problems in the lower digestive tract, a lower bowel cleanser herb combination, cascara sagrada, and chlorophyll have been used by many. Deodorant herbs like cloves, peppermint, and spearmint cover the odor. Acid breath is believed to be caused by a diabetes problem. Bad breath could also be caused by infection in the gums. Myrrh and white oak bark have been used to help correct this condition and to tighten the gums.

BALDNESS

Horsetail and capsicum are often used. Jojoba oil is recommended by many. Kelp and alfalfa are also popular. Sage is said to restore natural hair color. Using an herbal shampoo is also helpful. There is an herbal formula for hair, skin and nail problems.

BEDWETTING

Corn silk is a popular remedy. Peach bark is used by some practitioners. A kidney and bladder combination also is used.

BEE STINGS AND INSECT BITES

Black cohosh is taken internally. Comfrey is used as a poultice. Ice stops the swelling and rapid spread of the venom. Rose hips give high doses of cleansing vitamin C.

BLADDER

A kidney/bladder herb corrective combination is recommended for bladder and urinary problems. Peach bark, marshmallow, and corn silk are also used for these problems, as are white oak bark, golden seal, parsley, and dandelion.

BLOOD PRESSURE

To treat high blood pressure a heart herb combination, combination capsicum and garlic, a blood pressure combination, black cohosh, and gotu kola are used. To treat low blood pressure parsley and dandelion are popular remedies. Salt should not be used. The kidneys should be checked also.

BLOOD PURIFIER

One of the major benefits of herbs is to purify and clean the body. Special combinations have been traditionally used to purify the blood. One of the most popular is an herbal blood purifier combination which contains red clover, burdock, yellow dock, dandelion, licorice, chaparral, barberry, cascara sagrada, sarsaparilla, and yarrow. A person will only live as long as his blood stream is a fit place for his life force to reside. And through the blood stream every cell in the body is cleansed and fed.

BOILS

A soothing poultice herb combination works to relieve the boil. Burdock, chaparral, and dandelion are taken internally; also an herbal infection-fighting combination can be used.

BREASTS

A female corrective herb formula and saw palmetto are taken to enlarge the breasts. Chaparral and vitamins C and E are for treatment of lumps in the breast. It is very beneficial to stop using coffee, tea, alcohol, and chocolate.

BRONCHITIS

An asthma corrective herb combination is taken with vitamins C and A. Comfrey and fenugreek are also popular remedies. Lobelia, capsicum, and ginger are used to expel unwanted elements. Cleansing the bowels is beneficial.

BRUISES

Rose hips supply high doses of vitamin C which is believed to help healing. Comfrey is an herb especially useful for bruised or inflamed areas. It can be taken internally or applied as a poultice.

BURNS

Apply ice immediately to stop the internal burning. Aloe vera and a soothing poultice herb combination are mixed with vitamin E oil to form a paste to put on the burned area. Vitamins C and E are recommended. (See also the treatment of shock and comfrey.)

BURSITIS

A soreness in the joints is treated with an herbal combination for pain and a calcium rich herb combination primarily. An arthritis herb combination can also be used.

CHICKEN POX

Like most contagious diseases chicken pox must run its course. An herbal blood purifier combination helps clean the blood. An infection-fighting herb formula is used to fight the infection. Capsicum, garlic and golden seal are used sometimes. Many people bathe affected areas in soothing water that contains burdock, golden seal, and yellow dock. Lobelia is used to reduce the discomfort. Vitamin C or rose hips are believed to help clean the body. Vitamin A should be used since it fights all infections.

CHILDBIRTH

When labor starts take blue cohosh. An herb combination for nerves will help the woman relax. Since calcium is said to enable a person to withstand pain, a calcium rich combination helps. An herb combination for pain also helps. Lobelia is beneficial and could be taken if the above combinations do not contain it. Red raspberry should be used throughout pregnancy. A combination of herbs especially for the last 5 weeks of pregnancy is generally used.

CIRCULATION

Capsicum and garlic, blessed thistle, and vitamin E are common remedies. A heart herb combination is also a reputed help. Ginger is said to help circulation in the pelvic area and feet.

CLEANSING

Many different herbs and diets are recommended for cleansing. One popular cleansing herbal combination is an herbal blood purifier combination consisting of red clover, burdock, yellow dock, dandelion, licorice, chaparral, barberry, cascara sagrada, sarsaparilla, and yarrow. A lower bowel cleanser herb combination is another cleansing formula. A special herb tea made of red clover, chaparral, and spices is excellent for cleansing.

COLDS

A special herb combination for colds consisting of rose hip powder, camomile, yarrow, golden seal, peppermint, cayenne, lemon grass, sage, and myrrh has been traditionally formulated to treat colds. An infection-fighting herb formula combats infection. An herbal combination for lungs can be used to help with lung problems. Vitamin A, vitamin C, or rose hips are also popular aids.

COLIC

A catnip enema has proven successful for many babies. Mistletoe, catnip, and fennel are also popular.

COLON

Many different remedies are mentioned for treating the colon because of many different problems that can be found there. A lower bowel cleanser herb combination is used to eliminate accumulated waste. Cascara sagrada is a popular laxative. Slippery elm is purported to soothe the colon to promote healing. Herbal pumpkin or black walnut is used to expel parasites (start slow and build up because they can have a strong laxative effect.) A comfrey and pepsin combination is said to be able to clean the intestinal wall.

CONSTIPATION

As mentioned previously a lower bowel cleanser herb combination is formulated to cleanse the intestinal tract. Cascara sagrada is another popular laxative. Herbs high in potassium can also be used.

CONTAGIOUS DISEASES

Garlic and catnip enemas are often-used treatments. High vitamin A and C doses are also recommended. An infection fighting herb formula is used to combat infection. Garlic is a popular single herb. An herb combination formula for nerves is sometimes recommended to relieve pain and discomfort and to calm the nerves. Most contagious diseases are expected to run their course so the herbal remedies are primarily to help move things more quickly, help the body combat the infection more easily, or relieve the discomfort that accompanies the disease.

CONVULSIONS

Using an herb combination formula for nerves, a calcium rich herb combination, and B vitamins is a popular treatment. Lobelia and/or black cohosh are other herbs used to treat this situation. Nervine extracts can be used orally or in the ear.

COUGHS

Many herb formulas are known to aid this condition. After determining what causes the cough, remedies include an herb combination for colds, or an herbal combination for lungs, or a comfrey and fenugreek combination. Increased doses of vitamins A and C are advised. An asthma corrective herb combination proves helpful for many. There are also herbal cough syrups that are formulated to be expectorants.

CRAMPS

Menstrual cramps are usually treated with a female corrective herb formula and a calcium rich herb combination. Small amounts of black cohosh and dong quai aid many women. Ginger is sometimes used.

Muscle cramps are believed to be caused by a deficiency of calcium, so a calcium rich herb combination, alfalfa, and vitamins B and E are advised. Black cohosh can cause headache if too much is used.

CROUP

Garlic and catnip enemas are often used treatments. An infection fighting herb formula is used to combat infection. Garlic is a popular single herb. An herb combination formula for nerves is sometimes recommended to relieve pain and discomfort and to calm the nerves.

For the cough, remedies include an herb combination for colds, or an herbal combination for lungs, or a comfrey and fenugreek combination. Increased doses of vitamins A and C are advised. An asthma corrective herb combination also proves helpful for many people.

DANDRUFF

This unsightly problem is believed caused by disease, poor circulation, or nerve problems, or a combination of all these things. Each is treated with the appropriate herbal combination. Vitamin B, jojoba oil and herbal shampoos, capsicum, blessed thistle, and horsetail are acclaimed treatments.

DEPRESSION

For mental depression, an herbal "pick-up" combination, gotu kola, and kelp lessen or eliminate the problem, depending on severity and associated health factors.

DIABETES

Golden seal and licorice root are popular remedies. Dandelion root, blueberries, and raspberry are also often used. An herbal combination containing golden seal root, juniper berries, uva ursi leaves, huckleberry leaves, mullein leaves, comfrey root, yarrow herb, garlic bulb, capsicum, dandelion root, marshmallow root, buchu leaves, bistort root, and licorice root have been helpful.

DIAPER RASH

Slippery elm is mixed with vitamin E to form a soothing paste. Aloe vera can be used externally or taken internally as well. A popular home remedy is to sprinkle corn starch powder on the affected areas. Always bathe the baby with natural, organic soaps and wash the diapers in an organic, non-detergent soap to alleviate this problem.

DIARRHEA

It may seem strange, but a popular remedy is to use a lower bowel cleaner herb combination as a laxative to flush out undesirable elements that may be causing the diarrhea. Red raspberry is advised to stop diarrhea. Herbal pumpkin rids the body of unwanted parasites. Slippery elm helps soothe irritated areas. A calcium-rich herb combination and B vitamins are also popular. Psyllium adds bulk to bowel movement, thus slowing down the action.

DIGESTION

Papaya mint, safflower, capsicum, dandelion, hydrochloric acid (protein digestaid), and an herb combination for ulcers and sores solves the problems for many people. An anti-gas combination is sometimes helpful when stomach gas is a problem. Many times the problem is caused from a lack of food digestive enzymes. Herbs are super high in live enzymes.

DIZZINESS

Licorice or peppermint are often taken to improve low blood sugar which is a common cause of dizziness. Cleansing is also recommended to rid body of any unwanted parasites or toxins that could cause dizziness. Capsicum improves circulation.

DOUCHE

Golden seal mixed with water seems to be the most often utilized douche. Vinegar has also been used.

DRUG WITHDRAWAL

An extended program consists of taking camomile, golden seal, and licorice root the first week. The second week add the herbal "pick-up" combination, dandelion, and herb combination formula for nerves. High vitamin B and C doses are also part of the program.

DYSENTERY

Herbal pumpkin is recommended to eliminate undesirable parasites. Lower bowel cleanser herb combination also helps. Slippery elm soothes irritated membranes. Psyllium creates bulk.

EAR INFECTION (Earache)

Use extract of lobelia and vinegar and put drops into the affected ear. An herbal extract containing chickweed, black cohosh root, golden seal root, lobelia, scullcap, Brigham tea and licorice root is generally helpful.

ECZEMA

Popular treatments include purifying the blood with an herbal blood purifier combination, eating fresh fruits, keeping the bowels moving with lower bowel cleanser herb combination. Refrain from eating too many sweets and greasy foods. Vitamins A, C, and E and the herbs horsetail and chickweed have been used.

EMPHYSEMA

Comfrey and fenugreek combination is used with the herbal combination for lungs plus vitamins A, C, and E by many people affected with this disease.

ENDURANCE AND ENERGY

For the athletic type person, a fitness improvement herb combination is effective. The herbal "pick-up" combination is popular. Both work best when combined with a total program of exercise, good diet, and sufficient rest. Licorice, ginseng, and ho-shu-wu are popular. Other commonly used herbs include dandelion, yellow dock, gotu kola, kelp, bee pollen, and damiana.

EPILEPSY

Using herb combination formula for nerves, calcium rich herb combination, and B vitamins is a popular treatment. Lobelia and/or black cohosh and scullcap are other herbs used to treat this situation. Also clean the bowel and rid the body of parasites.

EYES

Vitamins A and C are taken with a specially formulated eye-wash combination of eyebright, golden seal, and bayberry.

FATIGUE

The herbal "pick-up" combination is popular. It works best when combined with a total program of exercise, good diet, and sufficient rest. Licorice, ginseng, and ho-shu-wu are popular. Other often used herbs include dandelion, yellow dock, gotu kola, kelp, bee pollen, and damiana.

FEET

For athlete's foot infection black walnut is used to eliminate the parasites. Foot odor is treated with chlorophyll. To improve toenails calcium rich herb combination proves helpful to many people. Callouses and corns are best treated by soaking the affected areas with salt water, then removing the excess build up on the skin surface.

FEVER

The infection fighting herb formula is popular for eliminating the infection. A catnip and garlic enema is an often used treatment. If ulcers or sores are the result of the fever, herb combination for ulcers and sores is popular. Other cleansing treatments include using lower bowel cleanser herb combination, herbal blood purifier combination, yarrow, vitamin A and vitamin C doses.

FEVER BLISTERS

The herb combination for ulcers and sores is popular. Aloe vera is recommended by many. Hydrochloric acid and pepsin are also used.

FINGERNAILS

The calcium rich herb combination builds stronger fingernails. Horsetail is also used.

FLU

The most obvious and utilized cure is herb combination for flu and vomiting. In addition herb combination for colds is taken with vitamins A and C.

FRACTURES

The calcium rich herb combination is utilized. Pain often accompanies fractures especially in the early stages so herbal combination for pain is popular in some cases. Comfrey is a popular single herb for this condition.

FRIGIDITY

Males: male hormone herb formula, ginseng, damiana, saw palmetto, sarsaparilla, licorice, and vitamin E.

Females: female corrective herb formula, damiana, ginseng, sage, fenugreek, vitamins E, dong-quai and licorice.

GALLBLADDER, GALLSTONES

A liver corrective herb formula is a most popular remedy for problems relating to the gallbladder. Cascara sagrada is said to have the ability to dissolve gallstones. Vitamins A and C are recommended.

GAS

Ginger, spearmint, and fennel are used to expel gas. An anti-gas combination is also effective. Protein digestaid may also be used.

GLANDS

A very effective herb remedy is a gland herb combination. For infection in the glands, infection-fighting herb formula works well.

GONORRHEA

Black walnut, infection fighting herb formula combinations, and vitamins A and C can be taken internally to combat the problem. A female corrective herb formula helps some women. Infection fighting teas and douches and suppositories can be used in the vaginal area.

GOITER

A thyroid corrective herb combination has been formulated to treat this problem. Kelp, black walnut, and white oak bark are also popular.

GOUT

A soreness in the joints is treated with herbal combination for pain and calcium rich herb combination primarily. Since it is very similar to arthritis, arthritis herb combination works well to help treat this problem and is often used with a special herbal formula. It is believed that calcium should be replenished with calcium rich herb combination. People over 45 often need a protein digestate (hydrochloric acid) to help digest some foods and calcium. Using an herbal blood purifier combination also is helpful. Safflower has been used very effectively. Using an herbal formula for cleansing the kidneys helps to remove the uric acid from the system. Also, cut down on acid producing foods.

GUMS

A paste made from white oak bark and myrrh is put onto cotton and held against the sore gums. An infection-fighting herb formula, and herb combination for ulcers and sores, and vitamin C doses are other treatments for this condition.

HAIR LOSS

Horsetail and capsicum are often used. A jojoba oil and herbal shampoo is recommended by many. Kelp and alfalfa are also popular.

HAIR COLOR

Sage and alfalfa are said to restore natural hair color.

HALITOSIS

Believed to be caused by problems in the lower digestive tract. The lower bowel cleanser herb combination, cascara sagrada, and chlorophyll have been used by many. Deodorant herbs like cloves, peppermint, and spearmint cover the odor. Acid breath is believed to be caused by a diabetes problem. Bad teeth and gum infections can be the cause of this.

HAY FEVER

A sinus herb combination used with the comfrey and fenugreek combination is beneficial to hay fever sufferers. Bee pollen, comfrey, and asthma corrective herb combination are often used. Also used are vitamins A, C, and E.

HEADACHE

The herbal combination for pain is a most popular remedy. Wood betony is a useful single herb. Calcium deficiency is believed to cause some headache problems, so the calcium rich herb combination is used. Cleaning the lower bowels with a catnip and garlic enema and with the lower bowel cleanser herb combination also relieves the problem.

HEART

The specially formulated herb combination helps many people. Vitamin E is often recommended in addition to the herb combination. Hawthorne berry is one of the most reputable herbs used for heart problems.

HEARTBURN

Papaya mint, hydrochloric acid (protein digestate), and herb combination for ulcers and sores solve the problem for many people.

HEMORRHAGE

Internal hemorrhage is treated with golden seal, capsicum, and white oak bark. External hemorrhage involves using capsicum and golden seal. Vaginal hemorrhage is treated with marshmallow, mistletoe, and capsicum. Comfrey also slows bleeding.

HEMORRHOIDS

One popular remedy is to take white oak bark internally or place the capsule in the rectum. Or make a tea and use a compress over the anus. One successful home remedy involves cutting a white potato into a suppository that is placed in the rectum. After inserting the potato for three successive days, a garlic suppository is used the fourth day. (Caution: if garlic is used without the potato treatment first, the tissue will probably become very sore and irritated.) it helps to put vitamin E on the potato. Also lecithin has been found to be very beneficial in chronic problems.

HERPES

Black walnut, golden seal, rose hips, and vitamin A are recommended. Herbal vaginal suppositories are very beneficial.

HIGH BLOOD PRESSURE

To treat high blood pressure the heart herb combination, a combination of capsicum and garlic, a blood pressure regulator combination, black cohosh, and gotu kola are used. To treat low blood pressure parsley and dandelion are popular remedies. Always check for kidney problems.

HOARSENESS

Licorice root, slippery elm, and sage tea have all been acclaimed as being soothing to the throat and helpful for treating this problem.

HORMONE IMBALANCE

Female: female corrective herb formula and black cohosh are proven remedies. Other popular single herbs include blessed thistle, damiana, sarsaparilla, and dong quai.

Male: male hormone herb formula is the most complete remedy; ginseng and sarsaparilla are popular single herbs that when taken together help balance male hormones.

HOT FLASHES

The female corrective herb formula and black cochosh are proven remedies. Other popular single herbs include blessed thistle, damiana, kelp, sarsaparilla, and dong quai. Vitamin E is also helpful. And in some cases ginseng helps.

HYPERACTIVITY

This is a difficult problem believed to be linked to sugar consumption, diet, and many other things. Herb combination formula for nerves, calcium rich herb combination, B vitamins, and licorice are often recommended.

HYPERGLYCEMIA

A special combination containing protein, capsicum, and garlic serves to correct the problems for many individuals. Licorice is also used. Fourteen different herbs in a combination have also been used.

HYPOGLYCEMIA

The special herb combination for hypoglycemia proves effective for many people. Treatments include rose hips, papaya mint, licorice root, and vitamins B, C, E and calcium.

INDIGESTION

Papaya mint, hydrochloric acid (protein digestate), and herb combination for ulcers and sores solve the problem for many people. An anti-gas combination is also popular for some stomach problems. The addition of a food enzyme formula helps. Capsicum, safflower and dandelion are also helpful.

INFECTION

The special infection-fighting herb formula is especially effective. Golden seal, rose hips, and vitamins A and C are recommended.

INFLAMMATION

The infection fighting herb formula works well to correct this problem. Rose hips supply high doses of vitamin C to help healing. Comfrey is an herb especially useful for bruised or inflamed areas. It can be taken internally or applied as a poultice.

INSECT BITES

Black cohosh is taken internally. Comfrey is used as a poultice. Ice stops the swelling and rapid spread of the venom. Rose hips give high doses of cleansing vitamin C.

INSOMNIA

A safe sleep herb formula seems to help many individuals with this problem. A calcium rich herb combination and herb combination formula for nerves are also often recommended. Camomile tea settles the stomach and helps some people relax and sleep.

ITCHING

Peppermint or ginger baths are helpful. Thyme mixed with vitamin E to form a paste when applied to the affected area reduces the itching. Corn starch, aloe vera gel, and tincture of black walnut are also popular remedies.

JAUNDICE

The liver corrective herb formula seems to help this condition since the disease is related to liver problems. Dandelion and vitamins A and C also help.

JOINTS

A soreness in the joints is treated with herbal combination for pain and calcium rich herb combination primarily. Safflower is also used.

KIDNEYS

The kidney/bladder herb corrective combination was formulated especially for bladder and urinary problems. Peach bark, marshmallow, and corn silk are also used for these problems, as are white oak bark, golden seal, parsley, and dandelion. Distilled water also helps.

LIVER

The liver corrective herb formula is formulated for all liver conditions. Dandelion and vitamins A and C also help.

LUMBAGO

A thorough cleansing of the lower intestinal area by using lower bowel cleanser herb combination helps many people. A garlic and catnip enema is another popular remedy. For the pain, herbal combination for pain, herb combination formula for nerves, and safe sleep herb formula are utilized. Building up calcium levels with calcium rich herb combination helps many sufferers, as 88 percent of all lower back pain is caused from lack of calcium.

LUNGS

Comfrey and fenugreek combination is used with the herbal combination for lungs plus vitamins A, C, and E by many people affected with lung disorders. Since infection may be involved, the infection fighting herb formula is recommended along with herbal combination for lungs, lobelia, and comfrey and fenugreek combination.

LYMPH (Swollen glands)

The special herbal formula along with infection fighting herb formula and vitamins A and C are beneficial for treating this problem. Golden seal, mullein and lobelia are also used by many people.

MEASLES

Garlic and catnip enemas are often used treatments. High vitamin A and C doses are also recommended. The infection-fighting herb formula is used to combat infection. Garlic is a popular single herb. The herb combination formula for nerves is sometimes recommended to relieve pain and discomfort and to calm the nerves. Most contagious diseases, like measles, are expected to run their course, so the herbal remedies are primarily to help move things more quickly, help the body combat the infection more easily, or help relieve the discomfort that accompanies the disease. Herbs are used to keep the bowels moving.

MENOPAUSE

During the time in a woman's life when changes are causing hormone imbalances, a female corrective herb combination, black cohosh, dong quai, vitamin E, and a calcium rich herb combination are the usual treatments.

MENSTRUATION

A female corrective herb combination with dong quai will help many menstrual problems. Black cohosh is used to relieve cramps. An herbal combination for pain helps relieve the discomfort of menstrual cramps. Blue cohosh is an often used remedy to regulate the menstrual cycle. Ginger helps to increase the flow of blood in the pelvic area.

MIGRAINE HEADACHE

The herbal combination for pain is a most popular remedy. Wood betony is a useful single herb. Calcium deficiency is believed to cause some headache problems, so the calcium rich herb combination is used. Cleaning the lower bowels with a catnip and garlic enema and with the lower bowels cleanser herb combination also relieves the problem for many people. Fenugreek and thyme combination helps in the area. Many times parasites are involved and can cause this problem.

MISCARRIAGE

Women who are more apt to have this problem have been helped by using red raspberry, female corrective herb formula, calcium rich herb combination, and vitamin E.

MORNING SICKNESS

This uncomfortable dilemma has been successfully treated with red raspberry, golden seal, female corrective herb formula, herb combination for ulcers and sores, herb combination for flu and vomiting, and doses of vitamin B.

MOUTH SORES

A paste made from white oak bark and myrrh is put onto cotton and held against sore gums. An infection fighting herb formula, an herb combination for ulcers and sores, and vitamin C doses are internal treatments for this condition.

MUCOUS MEMBRANES

The formulated herb combination for colds helps many people. The infection fighting herb formula, vitamins C and A, and the combination of comfrey and fenugreek also seem to help eliminate mucous problems where infections are involved.

MUMPS

This contagious disease must run its course. Vitamin C seems to help cleanse the body and aid the healing process. The special herbal formula along with infection-fighting herb formula and vitamin A are beneficial for treating this problem. Golden seal, mullein and lobelia are also used by many people.

NAUSEA

Red raspberry and herb combination for flu and vomiting are popular remedies. Peppermint, spearmint, and B complex vitamins seem to help some people.

NERVES

Especially helpful to many people is the herb combination formula for nerves. A safe sleep herb formula is another good remedy. Lobelia, calcium rich herb combination, and B vitamins are often recommended in addition to the previously mentioned herb combinations. Camomile tea before bed aids many people whose sleep is inhibited by nervous conditions.

NIGHTMARES

A safe sleep herb formula seems to help many individuals with this problem. A calcium rich herb combination and herb combination formula for nerves are also often recommended. Camomile tea settles the stomach and helps some people relax and sleep. Kelp is also used.

NURSING

The nutrient herb alfalfa is said to enrich the mother's milk with vitamins and minerals. If it is desired to increase the milk production, blessed thistle is recommended. For breast infections an infection fighting herb combination helps correct the problem. Sage decreases milk flow. Marshmallow increases the milk flow.

OBESITY

The specially formulated reducing aid herb formula is beneficial to many people to depress appetite. Chickweed and kelp are also popular. Fennel prevents hunger pains.

PAIN

The herbal combination for pain is a most popular remedy. Wood betony is a useful single herb. Calcium deficiency is believed to cause some pain problems so the calcium rich herb combination is used. Cleaning the lower bowels with a catnip and garlic enema and with the lower bowel cleanser herb combination also relieves the problem.

PANCREAS

A hypoglycemia/pancreas herb combination issued by many to treat problems resulting from pancreas problems. (See also Hyperglycemia, Hypoglycemia, and Diabetes.)

PARASITES

The most popular of all parasite remedies are garlic, black walnut or herbal pumpkin. A special herbal formula is also taken.

PERSPIRATION

For reducing excessive perspiration, heart herb combination, vitamin E, and calcium rich herb combination help many people. To eliminate perspiration odors, chlorophyll and alfalfa are popular. On those occasions when it is desirable to cause a person to perspire to help cleanse the body, a hot bath which has 2 tablespoons of ginger in the water works well. Leave the person in the water for 14 minutes. Upon leaving bath wrap in a warm blanket (perspiration is often so profuse that the person will completely soak one or two blankets.) also give a strong cup of yarrow tea upon leaving the bath.

PILES

One popular remedy is to take white oak bark internally or place the capsule in the rectum. A successful home remedy involves cutting a white potato into a suppository that is placed in the rectum. After inserting the potato for three successive days, a garlic suppository is used the fourth day. (Caution: if garlic is used without the potato treatment first, the tissue will probably become very sore and irritated.) It helps to put vitamin E on the potato. (See hemorrhoids.)

PITUITARY GLAND

Alfalfa and the herbal "pick-up" combination are popular remedies.

PNEUMONIA

Since infection is involved, the infection-fighting herb formula is recommended along with herbal combination for lungs, lobelia, and comfrey and fenugreek combination. Vitamins A and C are also recommended for this illness.

POISONING

Since this can be a very serious problem, contact your local poison control center as quickly as possible in emergency cases. Many kinds of poisoning require that the victim eliminate the poison by vomiting. Lobelia, when taken in large doses, is an excellent emetic that induces vomiting. For food poisoning and blood poisoning other remedies are used. For food poisoning, iceberg lettuce, large doses of lobelia, and large doses of vitamin C are used. For blood poisoning, the herbal blood purifier combination, large doses of vitamin C, and the infection fighting herb formula are administered. Never induce vomiting when there has been an intake of petroleum products.

POISON IVY AND POISON OAK

Peppermint or ginger baths are helpful. Thyme mixed with vitamin E to form a paste when applied to the affected area reduces the itching. Corn starch, aloe vera gel, and tincture of black walnut are also popular remedies. Also vitamin C is helpful for this condition.

PREGNANCY

Some herbs are highly recommended to be taken during pregnancy. Red raspberry helps with nausea and is a nutritive herb also. A calcium rich herb combination is recommended. Yellow dock helps for anemia. Vitamin E is highly recommended. And during the final weeks, a prenatal herb combination is taken by many herb users. (See also the other remedies associated with pregnancy such as morning sickness, childbirth, pain, etc.) It is almost universally recommended that black cohosh and pennyroyal NOT be taken by pregnant women, at least during the first seven months.

PROSTATE

The specially formulated prostate herb combination is excellent for problems with this organ. The male hormone herb formula aids many males also. The herbal pumpkin formula is used also.

PSORIASIS

Popular treatments include purifying the blood with an herbal blood purifier combination, eating fresh fruits, and keeping the bowels moving with lower bowel cleanser herb combination. See *The Little Vitamin and Mineral Encyclopedia* for which herbs are high in vitamins and minerals.

PYORRHEA

A paste made from white oak bark, comfrey and myrrh is put onto cotton and held against the sore gums. An infection-fighting herb formula, and herb combination for ulcers and sores and vitamin C doses are other treatments for this condition.

RHEUMATIC FEVER

The heart herb combination is very helpful. An infection fighting herb formula, vitamins A, C, and E, and the calcium rich herb combination are also often beneficial for this condition.

RHEUMATISM

An arthritis herb combination was created especially for treatment of this problem and is often used with a special herbal formula. Usually it is believed that calcium should be replenished with calcium rich herb combination. People over 45 often need a protein digestate (hydrochloric acid) to help digest some foods. Using an herbal blood purifier combination also is believed to be helpful. To digest calcium you must have enough hydrochloric acid.

RINGWORM

The most popular of all parasite remedies are black walnut, herbal pumpkin, and garlic. A special herbal formula is also taken.

SCARLET FEVER

An infection fighting formula; vitamins A, C, and E; and the calcium rich herb combinations are also often beneficial for this condition.

SENILITY

The herbal "pick-up" combination has helped many people to combat this condition. Blessed thistle and gotu kola are also popular.

SEX DESIRE

Males: male hormone herb formula, ginseng, damiana, saw palmetto, sarsaparilla, licorice, and vitamin E.

Female: female corrective herb formula, damiana, ginseng, sage, fenugreek, dong quai, vitamin E, and licorice.

SHINGLES

The herb combination formula for nerves, vitamin B complex, vitamin C, and lecithin help many sufferers. Applying a paste made with thyme and vitamin E oil to affected areas also brings comfort.

SHOCK

The herb combination formula for nerves, safe sleep herb formula, heart herb combination, capsicum, and lobelia along with vitamins B, C, and E are used to treat this special condition.

SINUS

A formulated herb combination for colds helps many people. An infection fighting herb formula, vitamins C and A, and the combination of comfrey and fenugreek also seem to help relieve sinus problems where infections are involved. Sniffing slippery elm or golden seal into the nasal passages aids many sufferers. (Bayberry can be inhaled too, but it is uncomfortable to many people.)

SKIN DISEASE

Popular treatments include purifying the blood with an herbal blood purifier combination, eating fresh fruits, keeping the bowels moving with a lower bowel cleanser herb combination. For itching, peppermint or ginger baths are helpful. A paste made by mixing thyme with vitamin E and applied to the skin reduces itching. Corn starch, aloe vera gel, a tincture of black walnut, vitamin A and chickweed are other good remedies for certain skin problems.

SLEEP

A safe sleep herb formula seems to help many individuals with this problem. A calcium rich herb combination and an herb combination formula for nerves are also often recommended. Camomile tea settles the stomach and helps some people relax and sleep.

SMOKING

Those wishing to quit smoking find help by using comfrey and fenugreek combination, camomile to settle and relax the stomach, licorice to build blood sugars and eliminate the craving caused by low blood sugar levels and herbal combination for lungs, a liver corrective herb formula, and high doses of vitamin C.

SORE THROAT

Capsicum works well along with an infection fighting herb formula, high doses of vitamin C, comfrey and fenugreek, and garlic and capsicum.

SPRAINS

Applying a soothing poultice herb combination many people find relief. An infection-fighting herb formula, a calcium rich herb combination, and vitamins C and E help to cleanse the area, aid healing, and fight the possibility of infection developing. Golden seal to reduce possible swelling is helpful.

SPLEEN

A liver corrective herb formula helps spleen problems since it works with the liver in normal functioning. Uva ursi is a single herb that helps also. High doses of vitamin C are recommended.

STERILITY

Balancing the sexual hormones proves beneficial to many situations. This is done by taking a male hormone herb formula, and a female corrective herb formula, and vitamin E.

STOMACH

Papaya mint, hydrochloric acid (protein digestate), and an herb combination for ulcers and sores are beneficial for many stomach problems.

SUNBURN AND SUNSTROKE

Cool the person and the skin. Aloe vera gel, cucumbers applied to the skin, vitamin E oil, and rose hips are recommended.

SWELLING

An infection fighting herb formula works well to correct this problem when infections are involved. Rose hips supply high doses of vitamin C which is believed to help healing. Comfrey is an herb especially useful for bruised, inflamed, or swollen areas—it can be taken internally or applied as a poultice. Also golden seal reduces swelling.

SWOLLEN GLANDS

A special herbal formula along with an infection fighting herb formula and vitamins A and C are beneficial for treating this problem. Golden seal, mullein and lobelia are also used by many people.

SYPHILIS

Black walnut, infection-fighting herb formula combinations, and vitamins A and C can be taken internally to combat the problem. A female corrective herb formula helps some women. Infection fighting teas and douches and suppositories can be applied by women to the vaginal area. Yarrow and yellow dock can be used also.

TEETH

To keep the teeth healthy and restore enamel, a calcium rich herb combination is taken internally, and the teeth are brushed with black walnut. For toothache, an herbal combination for pain, an herb combination formula for nerves, a safe sleep herb formula, and lobelia are all popular remedies. (Also see Gums.)

THROAT (Sore Throat)

Capsicum works well along with an infection fighting herb formula, high doses of vitamin C, comfrey and fenugreek, garlic and capsicum, and kelp.

THYROID

A thyroid corrective herb combination, kelp, and black walnut are often used to treat this condition.

TONSILLITIS

An infection-fighting herb formula helps the body combat infections that make the tonsils sore. A popular home remedy that shrinks and soothes the tonsils is to mix iodine and glycerin to an amber color and then paint the tonsils with a swab. Comfrey and fenugreek, and slippery elm help soothe the painful tonsils. Kelp and black walnut are also helpful.

TUMORS

A special herbal formula, a thorough blood cleansing using an herbal blood purifier combination, chaparral, and red clover are recommended treatments for this dangerous condition. A special tea made from red clover, chaparral, and other spices is very good. Vitamins A and C are valuable in helping this condition.

ULCERS

An herb combination for ulcers and sores helps this condition. A soothing poultice herb combination applied to the ulcer helps. Comfrey, a calcium rich herb combination, vitamins A and C, and aloe vera all taken internally are good for this condition. Vitamin E and aloe vera are often used for external application as well.

URINATION

A kidney/bladder herb corrective combination is formulated especially for bladder and urinary problems. Peach bark, marshmallow, and corn silk are also used for these problems, as are white oak bark, golden seal, parsley, dandelion, and alfalfa.

VAGINA

A female corrective herb formula taken internally helps many women. When infections are involved an infection fighting herb formula is recommended along with vitamins A, C, and E. Golden seal mixed with water is an often used douche. Herbal suppositories can be used. Vinegar douches are beneficial.

VARICOSE VEINS

The following herbs and herb combinations are taken internally to help this problem: white oak bark, a calcium rich herb combination, capsicum and garlic, alfalfa, and a heart herb combination. White oak bark is applied externally as well. Vitamins E, C and lecithin are also helpful.

VENEREAL DISEASE

Black walnut, yarrow, infection fighting herb formula combinations, and vitamins A and C can be taken internally to combat the problem. A female corrective herb formula helps some women. Infection fighting teas and douches and suppositories can be used in the vaginal area.

VITALITY

An herbal "pick-up" combination is popular. It works best when combined with a total program of exercise, good diet, sufficient rest. Licorice, ginseng, and ho-shou-wu are popular. Other often used herbs include dandelion, yellow dock, gotu kola, kelp, bee pollen, and damiana.

VOMITING

This uncomfortable dilemma has been successfully treated with red raspberry, an herb combination for flu and vomiting, and doses of vitamin B. Peppermint and spearmint seem to help some people.

WARTS

Vitamin A, garlic, chaparral, black walnut, and vitamin E are taken internally. Milkweed oil, castor oil, and black walnut can be applied externally.

WORMS

The most popular of all parasite remedies are black walnut or herbal pumpkin. A special herbal formula is also taken, as is garlic.

YEAST INFECTION

Take an infection fighting herb formula, high doses of vitamin C, capsicum and garlic, and vitamin A internally. A golden seal and garlic douche also helps. An herbal suppository can be used. Yogurt packs and eating yogurt is beneficial.

INDIVIDUAL HERBS

The effects of herbs may seem to be new to you, but they are not. Everyone has sneezed from a whiff of black pepper and cried while chopping an onion. Regretfully, millions of young people are daily experiencing the effects of some dangerous herbs such as the marijuana plant. *The Little Herb Encyclopedia Revised* lists only nonpoisonous, mild, safe herbs which you can use with confidence.

You will notice that we list the same herb as a treatment for several ailments. That's the beauty of herbs. They are amazingly versatile. Our bodies are uniquely individual, and what has achieved results for one may not be successful for you. There are many herbs that will work equally as well on one particular disease; some work better for one person than for another, thus many alternatives are given. It is not our intent, however, to diagnose or prescribe. We offer only advice that has worked for others.

An exciting new adventure in better health is about to unfold before you. Listed are the most common herbs that have the potential to immeasurably improve the quality of your life when combined with proper diet, rest, exercise, and a healthy mental attitude.

About Golden Seal

My daughter cut her heel on a nail on our carpeted stairstep. Infection soon set in with a red streak going up her leg. Golden seal powder and distilled water were used in paste form on the cut, then wrapped with plastic and left on overnight. The next morning the infection area was back to normal pink color, the red streak was gone and within two days a light scab had healed the cut. V.R.

About comfrey and the Calcium rich formula and poultice formula

My son had a motorcycle wreck and broke his leg, arm and cut his forehead quite badly. After the casts were put on his leg and arm and he was taken home, we started to give him a pain formula about 4 capsules every 1/2 hour and within 2 to 3 hours the pain had subsided. Then we give him 9 capsules of comfrey, 9 capsules of the calcium rich formula, and 9 capsules of the poultice formula every day. The doctors said it would take 12 weeks for the bones to heal and the casts to be taken off. With using the above, the casts were taken off in 5 weeks and he was playing football in 6 weeks. Hooray for herbs!!! V.R.

About Eye-wash formula

My mother was to have a cataract operation, but did not want to. We introduced her to the eye-wash formula and instructed her how to use it. She used the eye-wash externally in her eyes 2 to 3 times a day and took 4 capsules a day internally and within 3 weeks she could thread a needle and the doctor told her she did not need an operation! V.R.

About Capsicum and Garlic

I had very high blood pressure and had been doctoring for it for about 12 years to no avail. After using 6 capsules of capsicum and garlic a day for 2 weeks, my blood pressure was down to normal for the first time in 12 years. I also switched to the thyroid formula using about 6 a day instead of 4 grains of thyroid medicine a day, and shortly my thyroid problem was under control. M.P.

About Black Cohosh

I was having lumps reoccur in my breasts, and after having surgery 3 previous times, I did not desire this again. I used 6 Black Cohosh a day for 1 month then cut back to 2 to 3 capsules a day and within 2 months all the lumps were gone and the dryness in the vagina area was back to normal. M.Z.

Many different herb remedies have been successfully used to cure health problems. The uses of herbs explained in this book are common, often used herb applications, but they are not the only uses for those herbs. We believe that the descriptions give an accurate account of the most accepted uses for each herb listed. We offer only advice that has worked for others. It is not our intent to diagnose or prescribe.

ALFALFA

Alfalfa removes poisons or the effects of poison from the body and neutralizes acids. It is very high in minerals and vitamins. Alfalfa means "father of all foods" and contains all the vitamins and minerals known to man. It has the highest content of chlorophyll of any plant.

Allergies	Diabetes	Nausea
Anemia	Digestive disorders	Nutrients
Arthritis	Gout	Pituitary gland
Bladder	Inflammation	Rheumatism
Blood cleanser	Influenza	Stimulates appetite
Breath odor	Kidneys	Stomach
Bursitis	Lactation	Teeth
Colds	Liver	Ulcers
Cramps	Morning sickness	Whooping cough

ALOE VERA

Aloe vera is a tremendous healing agent when directly applied to abrasions, burns, and skin problems. It is very beneficial taken internally also.

Abrasions	Hemorrhoids	Skin eruptions
Athlete's foot	Poison ivy and oak	Regulates bowels
Burns	Promotes hair growth	Sores
Eczema	Psoriasis	Ulcers
Fever blisters		

ANGELICA

Angelica helps to relieve gas, promotes menstruation, and is generally invigorating and strengthening to the body.

Colds	Increases urination	Spleen
Colic	Liver	Stomach troubles
Fever	Lungs	Strengthens heart
Gas	Promotes menstruation	Ulcers
Heartburn		

BARBERRY

Barberry invigorates and strengthens the body. It increases bowel functions and stimulates appetite.

Anemia
Appetite stimulant
Bladder
Blood cleanser
Boils
Breath odor
Bright's disease
Cholera
Constipation
Diarrhea
Digestive disorders
Gallbladder

Gallstones
Heart
Heartburn
High blood pressure
Itching
Jaundice
Kidneys
Liver
Pyorrhea
Rheumatism
Ringworm
Skin disorders
Sore throat
Spleen
Typhoid fever

BAYBERRY

Bayberry is a stimulant which invigorates and strengthens the body; very desirable in treating any problem associated with the female organs. Used to check menstruation.

Bleeding
Bronchitis
Canker sores
Colon
Congested nose and sinuses
Cramps
Cuts
Diarrhea
Digestive disorders
Dysentery
Eyes
Fallen uterus
Fever
Flu
Gangrenous sores
Gargle
Headaches
Infection
Lumbago
Lungs
Menstruation supressant
Miscarriage
Mucus membranes
Scarlet fever
Sinus
Stomach
Sore throat
Thyroid
Ulcers
Vaginal discharge
Varicose veins

BEE POLLEN

Bee pollen is an energy food. It is highly esteemed by many nutritionists, who say it may be the only perfect food on the earth. It contains RNA and DNA.

Allergies
Asthma
Energy
Hay fever
Longevity

BISTORT

A very strong astringent which is most useful as a gargle and wash for sore mouth or gums. Used to increase the flow of urine and to stop bleeding.

Bedwetting	Expels worms	Skin eruptions
Cankers	Hemorrhage	Smallpox
Cholera	Insect stings	Snake bites
Diarrhea	Measles	Vaginal discharge
Dysentery	Regulates menstruation	Yellow jaundice

BLACK COHOSH

Black cohosh works directly on and calms the nervous system. It promotes menstruation and relieves menstrual cramps. Breaks up mucus and phelgm deposits. It also relieves or prevents spasms and causes perspiration.

Arthritis	Epilepsy	Nervous conditions
Asthma	Gallstones	Paralysis
Bee stings	Headaches	Pelvic disorders
Bowels	Heart palpitations	Poisonous bites
Blood cleanser	Hormone balance	Rheumatism
Bronchitis	High blood pressure	Skin disorders
Childbirth pains	Hysteria	Snake bites
Cholera	Insect bites	Spinal meningitis
Convulsions	Kidneys	Syphilis
Coughs	Liver	Thyroid
Diabetes	Lumbago	Typhoid fever
Diarrhea	Lungs	Uterine problems
Digestive disorders	Menopause	Whooping cough
Dropsy	Menstruation	Worms

BLACK WALNUT

Black walnut expels tapeworms and internal and external parasites. It is useful in treating tuberculosis and is high in iodine content.

Athlete's foot	Lactation	Teeth
Boils	Poison ivy	Throat
Colitis	Ringworm	Thyroid
Diarrhea	Skin diseases	Tuberculosis
Herpes	Syphilis	Vaginal discharge
Infections	Tapeworms	

BLESSED THISTLE

Blessed thistle acts to strengthen the heart and lungs and increase circulation to the brain. It stimulates milk production in nursing mothers.

Arthritis	Gallbladder	Kidneys
Constipation	Gas	Lactation
Cramps	Headaches	Lungs
Digestive disorders	Heart	Migraine headaches
Fever	Hormone balance	Vaginal discharge

BLUE COHOSH

Blue cohosh promotes the regulation of menstruation and stops false labor pains of childbirth. Causes profuse perspiration. Increases the flow of urine.

Bladder infection	Dropsy	Mucus
Blood cleanser	Epilepsy	Pain of childbirth
Blood pressure	Female problems	Rheumatism
Childbirth	Heart palpitations	Spasms
Colic	High blood pressure	Urine problems
Convulsions	Hysteria	Vaginal discharge
Cramps	Kidneys	Vaginitis
Diabetes	Menstruation	Whooping cough

BLUE VERVAIN

This herb has many medicinal properties. It is a quieting herb that helps calm coughing. It has cleansing properties in that it is a mild laxative and also causes perspiration. Expels worms when other herbs fail.

Asthma	Gallstones	Pneumonia
Bladder	Headache	Skin diseases
Colds	Insomnia	Smoking
Constipation	Kidney	Spleen
Convulsions	Measles	Sores
Expels worms	Menstruation	Stomach troubles
Fever	Mucus	Wounds
Female troubles	Nervousness	

BRIGHAM TEA

This desert herb is know for its ability to increase blood pressure and alleviate pain associated with arthritis and rheumatism. It increases nervousness and restlessness in some people.

Arthritis	Fever	Pain
Asthma	Headaches	Rheumatism
Blood pressure	Kidney	Scarlet fever
Blood purifier	Menstruation	Skin eruptions
Colds	Nose bleeds	Sinus

BUCHU

A good remedy for any problems of the urinary organs. Is especially soothing when there is pain while urinating. When taken warm, produces perspiration.

Bedwetting	Enlarged prostate gland	Urinary organs
Diabetes		Venereal disease
Dropsy	Pancreas	Weak bladder
	Rheumatism	

BUCKTHORN

Buckthorn is a bitter herb which expels impurities, increases the flow of urine, and produces nausea.

Constipation	Hemorrhoids	Skin disorders
Dropsy	Itching	Warts
Gallstones	Liver	Worms
Gout	Rheumatism	

BURDOCK

Burdock is an excellent blood purifier, cleansing and eliminating impurties from the blood very quickly. It also increases the flow of urine.

Allergies	Gout	Poison oak and ivy
Arthritis	Hair loss	Psoriasis
Bladder infections	Hay fever	Rheumatism
Blood cleanser	Hemorrhoids	Sciatic nerve
Boils	Inflammation	Skin disorders
Burns	Itching	Skin eruptions
Bursitis	Kidneys	Stomach disorders
Canker sores	Leprosy	Tuberculosis condition
Chicken pox	Liver problems	Ulcers
Gallbladder	Nervous conditions	Venereal diseases
Gallstones	Overweight	

CHAMOMILE

Chamomile, although a stimulant, is one of the finest nervine herbs that there is. It is strengthening and invigorating to the body. It is valuable in treating many ailments and addictions of all kinds. The herb is bitter.

Appetite stimulant	Dropsy	Muscle pain
Asthma	Drug withdrawal	Measles
Bladder troubles	Eyewash	Nervous conditions
Blood disorders	Expels worms	Regulates menstruation
Bronchitis	Gangrenous sores	Smoking
Callouses	Gas	Spleen
Colds	Headaches	Swelling
Colitis	Hemorrhage	Typhoid fever
Corns	Hemorrhoids	Toothaches
Cramps	Hysteria	Stomach
Dandruff	Indigestion	Yellow jaundice
Diverticulitis	Kidneys	

CAPSICUM

Capsicum is one of the most versatile herbs known to man. It can be applied directly to small open wounds to aid in stopping bleeding and healing. It may be used as a poultice for any inflammation and when taken internally will work to heal an ulcerated stomach; cleanses the circulatory system, reduces fevers, and purifies the blood. It has been used to help stop shock and aid in heart seizure.

Arteriosclerosis	Eyes	Pleurisy
Arthritis	Fever	Pyorrhea
Asthma	Gas	Rheumatism
Blood cleanser	Hay fever	Shock
Bronchitis	Heart	Skin problems
Chills	Hemorrhaging	Sore throat
Circulatory disorders	High blood pressure	Spasms
Colds	Indigestion	Spleen
Contagious diseases	Infection	Sprains
Convulsions	Inflammation	Stimulates appetite
Coughs	Kidneys	Tiredness
Cramps	Lockjaw	Ulcerated stomach
Cuts	Palsy	Varicose veins
Diabetes	Pancreas	Wounds
Digestive disorders	Paralysis	Yellow fever
External bleeding	Parkinson's disease	Yellow jaundice

CASCARA SAGRADA

Cascara sagrada is an excellent remedy for chronic constipation. It is a bitter tasting herb but is very invigorating to the body. It is called "sacred bark" by the Indians. It is not habit forming.

Chronic constipation	Gallbladder	Intestines
Colon	Gallstones	Liver problems
Cough	Gas	Nervous disorders
Croup	Hemorrhoids	Spleen
Digestive problems	Indigestion	Yellow jaundice

CEDAR BERRIES

(See Juniper)

CATNIP

Catnip relieves pain, prevents spasms, and calms the nerves. It also relieves gas from the bowels and causes perspiration. It is useful as an enema and is soothing and relaxing in general.

Acid stomach
Bronchitis
Colds
Colic
Convulsions
Diarrhea
Epilepsy
Expels worms
Fever
Gas
Headaches
Hypoglycemia
Hysteria
Inflammation
Insomnia
Kidney stones
Measles
Mental illness
Miscarriage
Morning sickness
Nervous conditions
Pain reliever
Shock
Skin problems
Smoking
Sleeplessness
Spasms
Stimulates menstruation
Stress
Tension
Uterine problems

CHAPARRAL

Many universities have tested chaparral and found it an aid in dissolving tumors and in cancer.

Acne
Allergies
Arthritis
Blood cleanser
Boils
Bursitis
Cancer
Cataracts
Cramps
Glaucoma
Hay fever
Kidneys
Overweight
Promotes hair growth
Prostate problems
Psoriasis
Rheumatism
Skin problems
Tumors
Warts

CHICKWEED

Chickweed is very soothing to any kind of internal inflammation. It is a good purifier in case of blood poisoning and can be applied externally to any skin problem including boils, burns, and sores in the mouth or throat.

Allergies
Appetite stimulant
Asthma
Blood poisoning
Boils
Breaks down fat
Bronchitis
Burns
Cancer
Circulatory problems
Colds
Constipation
Coughs
Deafness
Eyes
Hay fever
Hemorrhoids
Hoarseness
Itching
Impotency
Inflammed surfaces
Lungs
Peritonitis
Pleurisy
Psoriasis
Reducing aid
Rheumatism
Scalds
Skin diseases
Skin eruptions
Stomach problems
Swollen testes
Tumors
Ulcerated mouth-throat
Weight loss
Wounds

COMFREY

Comfrey (called the knitter and healer) stops hemorrhaging and bleeding and is particularly useful in treating bloody urine. It aids in soothing inflammation, and when used as a poultice can be applied to sore breasts, burns, wounds, swelling, and bites. It is a bone knitter.

Allergies
Anemia
Astham
Bladder problems
Blood cleanser
Bloody urine
Boils
Breasts
Bronchitis
Bruises
Burns
Bursitis
Colitis
Colon
Coughs
Cramps
Diabetes
Diarrhea
Digestive problems
Dysentery
Eczema
Emphysema
Female complaints
Fractures
Gallbladder
Gangrenous sores
Gout
Gum disease
Hay fever
Hemorrhaging
Hoarseness
Indigestion
Infections
Inflamed mucous membranes
Inflammation
Kidneys
Lungs
Menstruation
Pancreas
Psoriasis
Rheumatism
Sprains
Stomach
Swelling
Throat
Tiredness
Tonsillitis
Tuberculosis
Ulcerated kidneys, stomach, or bowels
Ulcers
Vaginal discharge
Wounds

CORN SILK

Corn silk is useful in controlling inflammation, relieving pain, and regulating urination.

- Bladder problems
- Kidney problems
- Painful urination
- Prevents bedwetting
- Prostate gland

DAMIANA

Damiana is especially good in treating female problems. It is also a mild aphrodisiac.

- Energy
- Female problems
- Hormonal balance
- Menopause
- Nervous conditions
- Parkinson's disease
- Prostate gland
- Reproductive organs
- Sexual impotency

DANDELION

Dandelion is especially good in treating diseases of the liver. It increases the flow of urine, acts as a gentle laxative, and is invigorating and strengthening to the body in general. It's high in organic sodium and vitamin A.

- Age spots
- Anemia
- Bladder
- Blood cleanser
- Boils
- Bowel inflammation
- Bronchitis
- Cancer
- Constipation
- Cramps
- Diabetes
- Digestive disorders
- Dropsy
- Eczema
- Female organs
- Fevers
- Gallbladder
- Gallstones
- Gas
- Gout
- Hemmorhage
- Hypoglycemia
- Indigestion
- Insomnia
- Kidneys
- Liver
- Low blood pressure
- Pancreas
- Psoriasis
- Rheumatism
- Scurvy
- Senility
- Skin eruption
- Skin problems
- Sores
- Spleen
- Stamina
- Stimulates appetite
- Tiredness
- Ulcers
- Urination
- Yellow jaundice

DON QUAI ROOT

Don Quai is used to treat almost every female gynecological ailment. A cosmetic side-affect, enlargement of the breast, has been reported by some women taking Don Quai. Sometimes it is called Tangkuei. Little or no fruit should be taken within 2 hours of taking Dong Quai for best results. It should not be taken during pregnancy. It is very high in vitamins E and B12.

Arthritic pain
Antispasmodic
Blood vessels
Blood purifier
Breasts, abcessed
Cancer
Cramps
Bronchitis
Fever
Heart
Hot flashes
Hypertension
Hypoglycemia
Insomnia
Laxative, mild
Lungs
Liver
Migraine headaches pain reliever
Menstruation regulates
Muscle tension
Prolapsed uterus
Pituitary
Plague
Releases retained placentia
Skin problems
Stomachache

ECHINACEA

An excellent blood cleanser. It works like penicillin in the body—with no side effects.

Bladder infection
Blood cleanser
Blood poisoning
Boils
Carbuncles
Diphtheria
Fever
Gangrenous conditions
Hemorrhage
Infections
Lymph glands
Mouth odor
Peritonitis
Poisonous bites and stings
Prostate glands
Skin eruptions
Sore gums
Sore throats
Syphilis
Tonsillitis
Wounds

EYEBRIGHT

As its name suggests, this herb is valuable in the treatment of all types of eye problems.

Allergies
Cataracts
Coughs
Diabetes
Digestive disorders
Ear ache
Glaucoma
Hay fever
Headache
Hoarseness
Vision aid

EUCALYPTUS

A fragrant herb which is useful in treating lung problems.

Bronchial tubes	Croup	Lung problems
Cancer	Fever	Paralysis

FALSE UNICORN

False unicorn is used to treat several female problems and especially for any trouble which may arise during pregnancy.

Diabetes	Miscarriage	Sterility
Fallen uterus	Promotes longevity	Strengthens uterus
Hemorrhage	Regulates menstruation	Vaginal discharge

FENNEL

Fennel is a spicy herb which relieves gas, increases the flow of urine, causes perspiration, and alleviates hunger pains.

Acid stomach	Eye wash	Migraine headaches
Bedwetting	Food poisoning	Overweight
Bronchitis	Gallbladder	Reducing aid
Colic	Gas	Rheumatism
Convulsions	Gout	Sinus trouble
Cough	Hoarseness	Snake bites
Cramps	Indigestion	Spleen
Curb appetite	Insect bites	Urination
Digestion	Liver	Yellow jaundice
Emphysema	Menstruation	

FENUGREEK

Fenugreek is very valuable when made into a poultice in treating all inflammations and wounds. Use as a gargle for sore throats. It also dissolves mucus.

Allergies	Emphysema	Lubricates intestines
Anemia	Eyes	Lungs
Bowel lubricant	Fever	Migraine headaches
Bronchitis	Hay fever	Mucus membranes
Bruises	Healing	Retained water
Coughs	Heartburn	Sore throat
Diabetes	Hoarseness	Vagina
Digestive disorders	Inflamed intestines	

GARLIC

Garlic works well in killing any harmful bacteria but will replace harmful bacteria with good bacteria. It also dissolves cholesterol.

Arteriosclerosis
Cancer
Contagious diseases
Coughs
Cramps
Diarrhea
Digestive disorders
Diverticulitis
Emphysema
Expels worms
Fever
Gallbladder
Gas
Heart
High blood pressure
Intestinal infection
Liver
Parasites
Prostate gland problms
Rabies
Rheumatism
Sinus problems
Ulcers
Warts
Yeast infections

GENTIAN

Gentian invigorates the entire system. It strengthens the digestive organs, increases circulation, and is good for female organs.

Appetite
Blood purifier
Circulation
Colds
Convulsions
Dysentery
Expels worms
Female organs
Fever
Gout
Jaundice
Liver
Menstruation
Relieves gas
Spleen
Strengthens digestive organs
Urination

GINGER

Ginger is a spicy herb which enhances the flavor of food, increases the secretion of saliva, and helps relieve gas. When it is taken hot, it also increases the flow of urine.

Chicken pox
Chronic bronchitis
Circulation
Colds
Colitis
Constipation
Contagious diseases
Cough
Cramps
Diarrhea
Digestive disorders
Gas
Gout
Headache
Hemorrhage
Indigestion
Influenza
Menstruation
Morning sickness
Mumps
Nausea
Paralysis
Pelvic circulation
Pneumonia
Shock
Sinus trouble
Sore throat
Stamina
Tongue paralysis
Vagina

GOLDEN SEAL

Golden Seal is of great value since it remedies a long list of ailments. It acts to invigorate and strengthen the body. It stimulates the bowels and acts to increase the flow of urine. It acts as an antiseptic and should be used to cleanse boils, wounds, and ulcers. Useful used as an eye wash. Kills poisons very effectively. Golden seal is very gritty so always strain it before putting it in the eye. It lowers blood sugar level; always use with licorice, if low blood sugar is a problem. It also reduces swelling.

Allergies
Antibiotic
Asthma
Breath freshener
Bladder problems
Bowel problems
Bright's disease
Bronchitis
Burns
Canker sores
Chicken pox
Circulation
Colds
Colitis
Constipation
Diabetes
Digestive disorders
Diphtheria
Eczema
Eye wash
Gallbladder
General cleanser
Gum diseases
Hay fever
Heart troubles
Hemorrhage
Hemorrhoids
Hoarseness
Infections
Inflammation
Influenza
Insulin
Intestines
Itching
Kidneys
Liver
Measles
Menstruation
Morning sickness
Mucus membranes
Nasal passages
Nausea
Nervous disorders
Overweight
Pancreas
Prostate gland
Psoriasis
Ringworm
Scarlet fever
Sinus trouble
Skin disorders
Small pox
Spleen
Syphilis
Thyroid
Typhoid fever
Tonsillitis
Ulcers
Urethra
Uterus
Vaginal discharge
Wounds

GOTU KOLA

Gotu kola is a "brain food" which promotes memory and energy and acts to prevent aging. It is effective in treating many mental problems. It is beneficial in the treatment of high blood pressure.

Age spots	Memory improvement	Prevents senility
Brain fatigue	Menopause	Promotes longevity
Clear voice	Mental troubles	Stamina
Depression	Nervous breakdown	Strengthens heart
High blood pressure	Pituitary	Vibrancy

HAWTHORN

Hawthorn has been used to aid in many kinds of heart problems. It is best to use it with capsicum and garlic.

Arteriosclerosis	Low blood pressure	Stamina
Arthritis	Menopause	Stress
Edema	Rheumatism	Vibrancy
Heart disease	Sleeplessness	

HO SHOU-WU

Ho Shou-Wu has been claimed by the Chinese to be an herb for longevity and restoration. Use large amounts and for a long period to get desired results as with all good tonic and restorative herbs.

Arthritis	Inflammation
Cancer (treat or prevent)	Post-partum and menstrual difficulties
Colds	Piles
Diarrhea	Tumors
Dilates blood vessels	Sedative
Fertility (promotes)	Ulcers
Heart	

HOPS

Hops serves to calm and soothe the nerves. It also relieves pain and reduces fever. It increases the flow of urine.

- Alcoholism
- Anxiety
- Bedwetting
- Coughs
- Delirium tremens
- Earaches
- Excessive sexual desires
- Fever
- Gonorrhea
- Headaches
- Hoarseness
- Indigestion
- Liver
- Menstruation
- Morning sickness
- Nervous conditions
- Overweight
- Restlessness
- Retained water
- Rheumatism
- Settles stomach
- Shock
- Sleeplessness
- Toothache
- Ulcers
- Worm expulsion
- Yellow jaundice

HORSERADISH

Horseradish improves the appetite and clears the sinuses

- Appetite
- Arthritis
- Asthma
- Bladder trouble
- Circulatory problems
- Cough
- Digestive disorders
- Expels worms
- Gout
- Hoarseness
- Liver
- Lungs
- Mucous membranes
- Nasal passages
- Rheumatism
- Retained water
- Sciatic nerve
- Spleen
- Wounds

HORSETAIL

Horsetail helps faciliate the use of calcium in the body. It has a high silica content, helps to strengthen fingernails, increases the flow of urine, and helps hold calkium in the body. It kills eggs of parasites, and disolves tumors.

- Bladder
- Convulsions
- Diabete
- Eyes
- Gas
- Hair loss
- Heart
- Hemorrhage
- Kidney stones
- Liver
- Nose bleed
- Mucus
- Nervous condition
- Parasites
- Skin problems
- Tumors
- Yellow jaundice

HYDRANGEA

Hydrangea increases the flow of urine. Will remove bladder stones and the pain caused by them. Will also relieve backache caused by kidney trouble.

Bladder stones	Dropsy	Paralysis
Bladder troubles	Kidney troubles	Scurvy
Chronic rheumatism		

IRISH MOSS

Irish moss is useful as a reducing aid. It is high in iodine.

Calcium retention	Mouth odor	Reducing aid
Cancer	Parathyroid	Thyroid trouble
Convulsions	Pneumonia	Yellow jaundice
Cough		

JUNIPER

Juniper works well to increase the flow of urine. It aids in preventing disease and destroys fungi.

Adrenal glands	Diabetes	Mucus
Allergies	Dog bites	Pancreas
Arthritis	Dropsy	Prostate gland
Bedwetting	Gargle	Retained water
Bladder	Gas	Scurvy
Blood	Gonorrhea	Snake bite
Boils	Hair loss	Stomach
Bright's disease	Hay fever	Throat
Contagious diseases	Hypoglycemia	Urinary trouble
Colic	Insect bites	Vaginal discharge
Coughs	Kidneys	Worms
Cramps	Leucorrhea	

KELP

Kelp aids in cleansing radiation from the body. It is high in iodine.

Adrenal glands	Energy	Morning sickness
Anemia	Goiter	Nails
Arteries	Hair loss	Pituitary
Bursitis	Infection	Pregnancy
Colitis	Kidneys	Prostate gland
Diabetes	Leg cramps	Psoriasis
Eczema	Menopause	Thyroid

LICORICE

Licorice contains a natural hormone that will replace cortisone. It raises blood sugar levels to normal.

Addison's disease	Cushing's disease	Sexual stimulant
Adrenal glands	Emphysema	Stamina
Age spots	Energy	Stomach
Arthritis	Female problems	Stress
Asthma	Hoarseness	Throat
Blood cleanser	Hyperglycemia	Ulcers
Bronchitis	Hypoglycemia	Vibrancy
Colds	Longevity	Voice
Constipation	Lungs	
Coughs	Menopause	

LOBELIA

Lobelia is a powerful relaxant, the strongest relaxant in the herb kingdom. Increases urine flow and perspiration. Large amount induces vomiting. Reduces fever. A poultice of lobelia is very soothing for inflammations, rheumatism, boils, etc.

Allergies	Food poisoning	Pain
Angina	Hay fever	Palsy
Arthritis	Headache	Pleurisy
Asthma	Heart	Pneumonia
Boils	Heart palpitations	Poison ivy
Bronchitis	Hepatitis	Promotes sleep
Bruises	Hoarseness	Relaxant
Bursitis	Hyperactivity	Rheumatic fever
Chicken pox	Hypoglycemia	Rheumatism
Congestion	Insect bites	Ring worm
Contagious diseases	Liver	Scarlet fever
Convulsions	Lockjaw	Shock
Coughs	Lungs	St. Vitus' dance
Croup	Meningitis	Toothache
Delirium	Migraine headaches	Tumors
Digestive disorders	Miscarriage	Wounds
Earache	Mucous membranes	Yellow jaundice
Emetic	Muscle motion	
Epilepsy	Nerve relaxant	

MANDRAKE

Mandrake regulates the bowels and is highly beneficial when used in chronic liver diseases. It also increases the flow of perspiration and urine. Dissolves and removes tumors. It is very strong and should be used with other herbs.

Colitis
Expels worms
Fever
Gallbladder
Gallstones
Liver
Regulates bowels
Uterine disease
Warts
Yellow jaundice

MARSHMALLOW

Marshmallow is very soothing for any sore or inflamed parts of the body.

Allergies
Asthma
Bedwetting
Bladder
Bleeding
Bloody urine
Bowels
Burn
Coughs
Diabetes
Diarrhea
Dysentery
Emphysema
Eyewash
Hay fever
Hoarseness
Inflammation
Influenza
Irritated vagina
Kidney-bleeding
Lactation
Lungs
Mucous membrane
Nervous condition
Pneumonia
Sore eyes
Sore throat
Voice loss

MISTLETOE

Mistletoe acts on the nervous system to quiet any excitement. It also relieves spasms. CAUTION: the berries are poisonous.

Cholera
Convulsions
Delirium
Epilepsy
Gallbladder
Heart troubles
Hemorrhage
High blood pressure
Hypoglycemia
Menstruation
Nervous conditions
Suppresses menstruation

MULLEIN

Mullein is soothing to any inflammation and relieves pain. It is useful in any chest ailment. Acts to relieve spasms and clears lungs.

Allergies	Dropsy	Pleurisy
Asthma	Dysentery	Pneumonia
Baby rashes	Earaches	Poison ivy
Bleeding bowels	Eye soreness	Sinus problems
Boils	Gargle	Skin disorders
Breathing problems	Glandular swelling	Swollen joints
Bronchitis	Hay fever	Throat soreness
Bruises	Hemorrhage	Tonsillitis
Constipation	Lung problems	Toothache
Coughs	Mucous membranes	Tumors
Croup	Mumps	Ulcers
Diarrhea	Pain reliever	Warts

MYRRH

Myrrh is a natural antiseptic which is good for bathing any kind of sores. Acts to stimulate and invigorate the body, and where necessary promotes menstruation. Where low blood sugar is a problem myrrh can be used in place of golden seal.

Antiseptic	Halitosis	Sores
Asthma	Hemorrhoids	Sore throat
Breath freshener	Indigestion	Thyroid
Boils	Infection	Toothache
Bronchial disease	Lung disease	Tuberculosis
Colitis	Menstruation	Ulcers
Colon	Mouth sores	Uterus
Coughs	Nervous conditions	Vaginal discharge
Cuts	Pyorrhea	Wounds
Diphtheria	Scarlet fever	
General healing	Shock	

OAT STRAW

Oat straw has a calming effect on the nerves. It helps hold calcium in the body and is very high in silica.

- Arthritis
- Bedwetting
- Bladder trouble
- Boils
- Bursitis
- Frostbite
- Gallbladder
- Gout
- Heart
- Improves appetite
- Kidney problems
- Liver disorders
- Lungs
- Nervous conditions
- Paralysis
- Rheumatism
- Skin
- Yellow jaundice

PAPAYA

Papaya has the highest enzyme content of any fruit known. Papain, one of its enzymes, is said to be a cancer preventer. Papaya contains all the enzymes needed for digestion of food and also relieves gas and sour stomach.

- Allergies
- Digestive disorders
- Diverticulitis
- Expels worms
- Hemorrhage
- Paralysis
- Relieves gas
- Sour stomach
- Wounds

PARSLEY

Parsley acts as a gentle laxative and increases the flow of urine. It is high in potassium.

- Allergies
- Arthritis
- Asthma
- Bedwetting
- Breath freshener
- Bruises
- Cancer
- Coughs
- Digestive disorders
- Dropsy
- Eyes
- Female troubles
- Fevers
- Gallbladder
- Gallstones
- Gout
- Hay fever
- Insect bites
- Kidney stones
- Liver
- Low blood pressure
- Menstruation
- Pituitary
- Prostate gland
- Retained water
- Spleen
- Stimulates appetite
- Stings
- Thyroid
- Tumors
- Urination problems
- Veneral diseases
- Yellow jaundice

PASSION FLOWER

Passion flower is very calming to the nerves; it's a mild sedative.

Alcoholism
Fever
Headaches
High blood pressure
Hysteria
Menopause
Nervous conditions
Restlessness
Sleeplessness

PEACH BARK

Peach bark relieves inflammation particularly in the bladder, acts as a mild laxative, and also has a soothing effect on the nerves.

Abdominal inflammation
Bladder
Bowel movements
Congestion
Expels worms
Hair loss
Jaundice
Morning sickness
Nausea
Nervous conditions
Sleeplessness
Stomach troubles
Uterine troubles
Whopping cough
Wounds
Yellow jaundice

PENNYROYAL

It is a spicy smelling herb that causes perspiration. Expels intestinal gas and has a sedative, calming effect on the nerves. Caution: this herb should not be taken by a pregnant woman.

Chest congestion
Colds
Colic
Convulsions.
Cramps
Fever
Gout
Headache
Insect bites
Intestinal gas
Intestinal pains
Itching
Jaundice
Lungs
Promotes menstruation
Mouth sores
Mucous
Nausea
Nervousness
Perspiration (promotes)
Skin diseases and irritations

PEPPERMINT

This spicy, pleasant smelling herb is used primarily to alleviate stomach and intestinal problems. It makes a stimulating tea, aids in digestion and is a catalyst for other herbs.

Appetite
Boils
Chills
Cholera
Cleanser
Colds
Colic
Colon trouble
Coughs
Cramps (stomach)
Diarrhea
Digestion
Diverticulitis
Dizziness
Fevers
Flu
Gallbladder
Gas
Headache
Heart palpitations
Hysteria
Insomnia
Itching
Liver
Measles
Menstruation
Migraine headaches
Morning sickness
Muscle spasms
Nausea
Nerves
Nightmares
Rheumatism
Shingles
Smoking
Stimulant
Vomiting

PLANTAIN

This beneficial herb aids healing. It fights infection and checks bleeding. Promotes urine secretion. An excellent poultice for wounds, burns, and skin irritations.

Bladder
Bleeding
Boils
Burns
Diarrhea
Douche
Eyes
Frigidity
Hemorrhage
Hemorrhoids
Insect bites
Itching
Kidneys
Leucorrhea
Lower back pain
Lungs
Menstruation
Piles
Poisoning
Scalds
Sores
Skin irritations
Thrush
Toothache
Tumors
Ulcers
Vagina
Venereal disease
Wounds

PLEURISY ROOT

It is useful for calming spasms and relaxing the body. It increases urine secretions and perspiration. It checks coughing and gas.

Asthma
Bronchitis
Circulation
Colds
Contagious diseases
Coughs
Dysentery
Fever
Flu
Kidneys
Lungs
Measles
Menstruation
Mucus
Perspiration
Pleurisy
Pneumonia
Rheumatic fever
Scarlet fever
Typhus
Water retention

POKE ROOT

The herb comes from the root of the poke plant. It has general healing qualities for the whole body. It is especially good for dissolving tumors, cleansing boils and sores, and for fighting some veneral diseases.

Arthritis
Boils
Breasts
Cancer
Glands
Goiter
Gums
Hemorrhoids
Infection
Inflammation
Itching
Kidney
Laxative
Liver
Lumbago
Lymph glands
Pain
Parasites
Rheumatism
Skin disease
Sores
Thyroid
Tremors
Tumors
Veneral disease

PRIMROSE SEED

Primrose oil is acquired by extracting it from the seeds of this plant. It is high in vitamins C, B6 and niacin and zinc.

Alcoholism
Anterial lateral sclerosis
Arthritis
Breast cysts
Cancer
Heart
High blood pressure
Inflammation
Hangovers
Mental illness
Multiple sclerosis
Nerves
Obesity
Reduces cholesterol
Schizophrenia
Skin problems
Stimulates immune system
Weight loss

PSYLLIUM

Psyllium is an excellent cleanser for the intestines and colon. It also acts to lubricate, moisten, and heal the intestinal tract. Psyllium is very moisture absorbing; always take plenty of water with it since it expands, and always use an herbal laxative with it

Colitis	Diverticulitis	Intestinal tract
Colon cleanser	Hemorrhage	Ulcers
Constipation		

QUEEN OF THE MEADOW (also called Gravel Root)

This herb increases urine flow. It has astringent qualities. It is known as a mild relaxant. It works especially well in combination with other herbs.

Diabetes	Nervous conditions	Urination
Dropsy	Prostate	Uterus
Gravel	Relaxant	Vagina
Kidneys	Rheumatism	Water retention
Lumbago		

RED CLOVER

Red clover is an excellent blood purifier. It can be used extensively in the treatment of cancer. Bathe any sores in the tea.

Blood cleanser	Nervous conditions	Skin disorders
Boils	Psoriasis	Spasms
Bronchial troubles	Rheumatism	Syphilis
Cancer	Scarlet fever	Whooping cough

RED RASPBERRY

Raspberry helps to promote painless and bloodless childbirth. It helps quiet nausea and acts to stop diarrhea, especially in children. Take throughout pregnancy.

Bronchitis	Fever	Nervous conditions
Canker sores	Flu	Nursing pain
Colds	Influenza	Painless childbirth
Constipation	Lactation	Pregnancy
Coughs	Measles	Rheumatism
Diabetes	Menstruation	Stomach
Diarrhea	Miscarriage	Strengthens uterine walls
Digestive disorders	Morning sickness	Throat
Dysentery	Mouth sores	Ulcers
Eyewash	Mucous membranes	Vaginal discharge
Female organs	Nausea	

REDMOND CLAY

Redmond clay is particularly useful in treating skin disorders. It also cleanses worms from the intestinal tract. Always take with an herbal laxative.

Bee stings	Skin disorders
Expels worms (intestinal tract)	Skin eruptions

RHUBARB

It is a gentle laxative that strengthens the gastrointestinal tract as well. It tones and tightens body tissues.

Colon	Headache	Liver
Constipation	Gallbladder	Stomach
Diarrhea		

ROSE HIPS

Rose hips is used to fight infection and curb stress. It is the highest herb in vitamin C content and contains the entire vitamin C complex.

Arteriosclerosis	Contagious diseases	Kidneys
Bites	Emphysema	Stings
Bruises	Fever	Stress
Circulation	Heart	Yellow jaundice
Colds	Infections	

SAFFLOWER

Safflower helps to increase the bowel functions and increase the flow of urine. It produces perspiration and also can be used to promote menstruation and relieve gas. It is the food the body uses for production of hydrochloric acid and neutralize uric and lactic acid.

Acid stomach	Heartburn	Retained water
Arthritis	Hyperglycemia	Scarlet fever
Bronchitis	Hypoglycemia	Skin diseases
Digestive disorders	Measles	Stimulates appetite
Gas	Menstruation	
Gout	Psoriasis	

SAGE

Sage is a popular seasoning for dressing, soups, and so on; it is also useful to quiet the nerves, relieve spasms, produce perspiration, and expel worms from children and adults. It also helps return hair to its original color.

Bites	Headache	Palsy
Bleeding wounds	Hoarseness	Pneumonia
Bronchitis	Influenza	Relieves gas
Circulation	Kidney	Regulates sex desire
Cleanse sores	Lactation	Skin disorders
Dandruff	Liver	Sore throat
Diarrhea	Lungs	Stomach troubles
Digestive disorders	Menstruation	Tonsillitis
Expels worms	Morning sickness	Ulcerated mouth and throat
Fever	Nausea	
Gas	Nervous conditions	Voice loss
Hair tonic	Night sweat	

SARSAPARILLA

Sarsaparilla is an excellent antidote for poison. It relieves inflammation and gas and will increase the flow of urine. Also used to promote perspiration.

Age spots	Heartburn	Relieve gas
Blood cleanser	Hormone regulation	Ringworm
Boils	Male hormones	Rheumatism
Colds	Menopause	Sexual problems
Eyewash	Mucus buildup	Skin eruptions
Fever	Poison antidote	Veneral disease
Gout	Psoriasis	

SAW PALMETTO

Saw palmetto is quieting to the nerves and acts as an antiseptic. It is claimed to increase the size of the breasts of women of childbearing age.

Alcoholism
Asthma
Bladder
Diabetes
Bright's disease
Colds
Enlarges small breasts
Hormone regulation
Mucus discharge of head area
Promotes weight gain
Prostate gland
Reproductive organs
Throat troubles
Whooping cough

SCULLCAP

Scullcap acts on the nervous system, calming nervous excitement. It relieves spasms and increases the flow of urine. It helps rebuild nerve endings in the brain.

Alcoholism
Convulsions
Epilepsy
Fits
Headaches
Hysteria
Hypoglycemia
Indigestion
Mental illness
Migraine headaches
Mumps
Nervous conditions
Palsy
Paralysis
Poisonous bites
Rabies
Rheumatism
Sleeplessness
Smoking
Strengthens heart spasms
Supresses sex desire

SIBERIAN GINSENG

Ginseng helps to strengthen and tone the stomach and relieves inflammation. It contains male and female hormones.

Age spots
Aging
Asthma
Bleeding
Cancer
Chest trouble
Colds
Constipation
Coughs
Digestive disorders
Lung problems
Impotency
Promotes longevity
Prostate gland
Skin eruptions
Stimulates appetite
Stomach problems
Urinary tract inflammation

SLIPPERY ELM

Slippery elm is soothing to any inflamed or irritated areas.

Asthma
Baby rashes
Bladder
Boils
Bowels
Bronchitis
Burns
Cancer
Colitis
Colon
Constipation
Cough
Cramps
Cystitis
Diabetes
Diarrhea
Digestive disorders
Diverticulitis
Dysentery
Eczema
Expels worms
Eyes
Fever
Gangrenous wounds
Gas
Hay fever
Hemorrhage
Hoarseness
Inflammation
Influenza
Kidneys
Lungs
Reproductive organs
Stimulates sex desire
Stomach
Strong perspiration
Tonsillitis
Ulcers
Urinary tract diseases
Vaginal discharge
Venereal disease sores
Wounds
Yellow jaundice

SPEARMINT

This spicy herb is very similar to peppermint in medicinal properties. But, unlike peppermint, it should never be boiled to make a tea. It's calming and soothing to the stomach and intestine. It increases the circulation in the stomach.

Appetite
Bladder
Colon trouble
Diarrhea
Digestion
Gas
Headache
Hemorrhoids
Insomnia
Kidneys
Morning sickness
Nausea
Nervousness
Piles
Urination
Vomiting

SPIRULINA

Spirulina is an algae containing 65 to 70% protein, 26 times the calcium of milk, lots of phosphorus and niacin and is far more nutritious than any known food. Highest source of B12. Keeps in storage without vacuum pack for years without any special preservatives.

Arthritis
Blood cleansing
Colon cleanser
Energy
Hypoglycemia
Rejuvenation
Weight reduction

SQUAWVINE

Noted for its ability to ease childbirth, it was a popular Indian remedy, thus the name squawvine. It is a healing herb that promotes urination as well as tones body tissue.

Bladder	Hemorrhoids	Pain
Childbirth	Inflammation	Piles
Eyewash	Menstruation	Urination
Gravel	Nerves	Wounds

ST. JOHNSWORT

St. Johnswort acts to dissolve and remove tumors. It calms the nerves and increases the flow of urine. It is an excellent blood cleanser.

Afterbirth pains	Gout	Regulates menstruation
Anemia	Heart	Sciatica
Bedwetting	Hemorrhage	Tumors
Blood cleanser	Hysteria	Urination
Boils	Lower back spasms	Uterine troubles
Coughs	Lungs	Wounds
Diarrhea	Nervous condition	Yellow jaundice
Dysentery	Palsy	

TAHEEBO (PAU D'ARCO)

Taheebo is a South American herb obtained from the bark of the Red Lapacho tree and used by the Indians curing cancer and many other ailments.

Anemia	Gonorrhea	Psoriasis
Antibiotic	Hemorrhages	Rheumatism
Arteriosclerosis	Hodgkins disease	Ringworm
Asthma	Inflammation of genital system	Scabies
Bronchitis	Leukemia	Skin disease
Cancer (all types)	Leukorrhea	Syphilis
Relieves pain of cancer	Lupus	Sequels
Colitis	Osteomyletis	Ulceration of intestines
Cystitis	Paralysis of eyelids	Ulcers (gastric & duodenal)
Diabetis	Parkinson's disease	Varicose (ulcers)
Eczema	Polyps (intestinal vesical)	
External sores		
Gastritis		

THYME

Thyme helps to dissolve and remove tumors. It relieves spasms and gas; encourages menstruation.

Asthma
Colic
Cramps
Diarrhea
Digestive disorders
Fever
Gas
Headache
Heartburn
Influenza
Lungs
Migraine headaches
Mucus
Mucous membranes
Nervous condition
Nightmares
Promotes menstruation
Shingles
Tumors
Weak stomach
Whooping cough

UVA URSI

Uva ursi is invigorating and strengthening to the body. It also increases the flow of urine and cleanses the spleen.

Bedwetting
Bladder
Bright's disease
Bronchitis
Cyctitis
Diabetes
Digestive disorders
Venereal diseases
Dysentery
Female troubles
Hemorrhoids
Kidneys
Liver
Lumbago
Mucous membranes
Overweight
Pancreas
Prostate gland
Regulates menstruation
Spleen
Vaginal discharge

VALERIAN ROOT

Valerian root works to calm the nerves. Also relieves pain and spasms.

Afterbirth pains
Blood pressure
Colds
Colic
Contagious diseases
Convulsions
Digestive disorders
Fever
Gas
Headache
Heart palpitations
Hypoglycemia
Hysteria
Measles
Migraine headaches
Muscle spasms
Nervous conditions
Pain reliever
Palsy
Paralysis
Promotes menstruation
Restlessness
Scarlet fever
Shock
Skin eruptions
Sleeplessness
Ulcerated stomach

VERVAIN

See Blue Vervain.

WHITE OAK BARK

White oak bark stops bleeding in the stomach, liver, and bowels. It increases the flow of urine and acts as a good antiseptic and astringent. It is useful for varicose veins and hemmorhoids.

Bleeding stomach, liver, bowels
Bruises
Expels pinworms
Fever
Fever blisters
Gallstones
Goiter
Hemorrhage
Hemorrhoids
Kidney stones
Liver
Menstruation
Mouth sores
Pyorrhea
Skin diseases
Skin eruptions
Spleen
Thrush
Tonsillitis
Toothache
Tumors
Ulcerated bladder
Ulcers
Urination
Vaginal discharge
Varicose veins
Yellow jaundice

WILD YAM

Wild yam is known for its ability to soothe nerves, treat liver related problems, and help with general pains during pregnancy.

Cramps
Liver
Menstruation
Morning sickness
Nausea
Nerves
Pain
Prevent miscarriage
Rheumatism
Stomach
Spasms
Ulcers
Urination

WITCH HAZEL

Witch hazel has a slight sedative property when taken internally. It is very useful for treating inflamed or irritated sensitive tissues like mucous membranes.

Circulation
Diarrhea
Eyes
Gums
Hemorrhage
Inflammation
Mucous membranes
Nose bleed
Piles
Sinus
Sores
Sore throat
Varicose veins
Venereal disease
Wounds

WOOD BETONY

Wood betony is calming to the nerves. It acts as gentle laxative. It can prevent or cure scurvy and tones the stomach. It is a vascular dialator, and excellent for headaches.

Asthma
Bedwetting
Bladder
Bronchitis
Colds
Colic
Convulsions
Consumption (T.B.)
Cough
Delirium
Digestive disorders
Dropsy
Expels worms
Gout
Headache
Heartburn
Inflammation
Kidneys
Liver obstructions
Menstruation
Mental illness
Migraine headaches
Nervous conditions
Pain reliever
Palsy
Parkinson's disease
Poisonous bites
Scurvy
Spleen obstructions
Stomach cramps
Tonsillitis
Varicose veins
Wounds
Yellow jaundice

WORMWOOD

Wormwood is an invigorating, aromatic herb that has antiseptic qualities and is capable of expelling worms and other parasites.

Appetite
Arthritis
Boils
Bruises
Childbirth
Circulation
Diarrhea
Expels worms
Fever
Gallbladder
Gas
Indigestion
Liver
Lower back pain
Parasites
Sprains
Swelling
Stomach

YARROW

Yarrow can be used to control bleeding from the lungs and bowels. Regulates many urination problems. It soothes and heals mucous membranes and is effective against fever and flu.

Arthritis	Diabetes	Lungs
Bladder	Diarrhea	Measles
Bleeding lungs and bowels	Dysentery	Menstruation
	Ear infections	Mucous membranes
Blood cleanser	Female troubles	Night sweats
Bright's disease	Fever	Pleurisy
Bursitis	Gas	Smallpox
Chicken pox	Hair loss	Spleen
Colds	Hemorrhoids	Typhoid fever
Colon	Influenza	Uterus
Congestion	Kidneys	Vaginal discharge
Contagious diseases	Liver	Yellow jaundice

YELLOW DOCK

Yellow dock strengthens and tones the entire system. It is very valuable for cleansing the blood, boils, ulcers, and wounds. It is high in iron.

Acne	Glandular tumors	Skin eruptions
Anemia	Itching	Spleen
Bladder	Leprosy	Stamina
Bleeding	Liver	Swelling
Blood cleanser	Paralysis	Syphilis
Boils	Pituitary	Tiredness
Cancer	Poison ivy, oak	Ulcers
Chicken pox	Psoriasis	Veneral disease
Ear infection	Runny ears	Vibrancy
Eyes	Scarlet fever	Wounds
Gallbladder	Scurvy	

YUCCA

Primarily used for arthritic and rheumatoid conditions.

Blood Purifier
Detoxifier
Skin Problems

HERB COMBINATIONS

Herbalists who have devoted their lives to understanding the medicinal properties of nature's healers believe that many herbs work best when used in combination with other herbs. This is because some act as catalysts, encouraging other herbs to work more effectively.

Seldom can we isolate one single deficiency or excess as the cause of a medical problem. Illness is generally the result of an accumulation of ailments, thus herbal combinations are often the most effective treatment.

The Little Herb Encyclopedia Revised takes the guesswork out of herb combinations by suggesting some of the most beneficial herbal formulas. As with single herbs, time and experimentation will determine which work best for your body. Depending upon where we live and the foods our bodies are accustomed to, some herbs and herb combinations will have a rapid and dramatic effect. Others will work more slowly.

Realize that some contagious diseases must run their course, and although herbal formulas may be helpful in treating the symptoms, the body must cure itself.

The herb combinations listed here and their uses are only suggested for your benefit. *The Little Herb Ecyclopedia Revised* does not attempt to diagnose or prescribe.

Stories abound telling of the tremendous benefits people have received from using herbs.

ANTI-GAS

papaya fruit, ginger, peppermint, wild yam, fennel, lobelia, spearmint, and catnip.

Recommended Dosage and Use: two capsules at mealtimes to relieve gas problems related to digestion and indigestion.

ANTI-DEPRESSION/"PICK-UP"

gotu kola, ginseng, capsicum.

Recommended Dosage and Use: two capsules as needed — this herb combination is designed to be used when needed to combat depression. Can be used anytime for energy; also promotes longevity.

ARTHRITIS (or Rheumatism)

yucca, bromelain, alfalfa, comfrey, chaparral, burdock, black cohosh, yarrow, capsicum, lobelia, centaury.

Recommended Dosage and Use: three capsules at mealtime to aid arthritis sufferers.

ARTHRITIS (or Rheumatism)

hydrangea, Brigham tea, chaparral, yucca, black cochosh, capsicum, black walnut, valerian, sarsaparilla, lobelia, scullcap, burdock, wild lettuce, and wormwood.

Recommended Dosage and Use: two capsules at mealtimes to maintain health and relieve arthritis suffering.

ASTHMA

blessed thistle, black cohosh, scullcap, pleurisy root.

Recommended Dosage and Use: take two or more capsules per meal to combat asthma attacks and allergies. Also used for other bronchial and lung area problems.

BLOOD PRESSURE EQUALIZING

garlic, capsicum, parsley, ginger, Siberian ginseng, and golden seal.

Recommended Dosage and Use: two capsules at mealtimes to maintain proper pressure levels.

BLOOD PRESSURE REDUCER

capsicum, garlic.

Recommended Dosage and Use: two capsules with each meal to lower blood pressure levels.

BLOOD PURIFIER

red clover, burdock, yellow dock, dandelion, licorice, chaparral, barberry, cascara sagrada, sarsaparilla, yarrow.

Recommended Dosage and Use: two capsules at mealtimes to maintain health by purifying the blood, combating contagious diseases and for general blood cleansing.

BLOOD PURIFIER (extra strength)

licorice, red clover, sarsaparilla, cascara sagrada, Oregon grape, chaparral, burdock, buckthorn, prickly ash, peach, and stillingia.

Recommended Dosage and Use: two capsules with two meals each day to maintain a healthy body.

CALCIUM RICH

Irish moss, alfalfa, horsetail, comfrey, lobelia, oatstraw.

Recommended Dosage and Use: two capsules with each meal to increase calcium levels in the body. Referred to as the knitter, the healer.

CALCIUM

comfrey, horsetail, oat straw, and lobelia.

Recommended Dosage and Use: two capsules with morning and evening meals each day to replenish low calcium levels.

CLEANSER/PAIN RELIEF

gentian, catnip, golden seal, barberry bark, myrrh gum, yellow dock, Irish moss, comfrey, mandrake, fenugreek, chickweed, black walnut, dandelion, safflower, St. Johnswort, echinacea, cyani flowers.

Recommended Dosage and Use: two capsules before each meal daily to maintain health. This special formula herb combination has wide application but is most often used as a blood cleanser, for arthritis correction, and as a pain reliever such as for headaches.

COLITIS

comfrey, marshmallow, slippery elm, ginger, wild yam, and lobelia.

Recommended Dosage and Use: two capsules with three meals each day to maintain a healthy colon. All citrus should be eliminated while this condition exists. Also cold liquids should not be taken.

COLD REMEDY

rose hips, chamomile, slippery elm, yarrow, golden seal, peppermint, capsicum, lemon grass, sage, myrrh.

Recommended Dosage and Use: three capsuies every three hours to combat common cold symptoms. This combination is a widely used and very popular herb remedy.

COLDS/FLU

bayberry, ginger, white pine, capsicum, and cloves.

Recommended Dosage and Use: two capsules with three meals each day to combat colds and flu. This is Jethro Kloss's famous herbal composition powder.

COLDS/VIRUS

garlic, rose hips, rosemary, parsley, and watercress.

Recommended Dosage and Use: two capsules three times per day at mealtime to combat viral cold infections.

DIGESTIVE AID

papaya, peppermint leaves, enzymes with HCL and pepsin.

Recommended Dosage and Use: one or two capsules at mealtime to aid digestion and calm the stomach.

EYEWASH

golden seal, bayberry, eyebright.

Recommended Dosage and Use: mix one capsule in warm water, let stand for approximately fifteen minutes until water is cool, strain, use the liquid to bathe the eyes 2-to-3 times daily or as desired. May be used internationally for strengthening the eye, 2-to-3 times daily.

FASTING

licorice, beet root, hawthorne berry, and fennel seeds. Controls hunger and helps keep blood sugar level at normal while on a fast.

Recommended Dosage and Use: 2 capsules at each meal.

FEMALE CORRECTIVE

red raspberry, golden seal, queen of the meadow, marshmallow, blessed thistle, lobelia, capsicum, black cohosh, and ginger root.

Recommended Dosage and Use: two capsules at mealtime to correct many different female problems by balancing hormones. After things are normalized take only two capsules daily. Relieves severe cramps.

FEMALE CORRECTIVE WITH DONG QUAI

red raspberry leaves, dong quai root, ginger, licorice root, black cohosh root, queen of the meadow, blessed thistle, marshmallow, ginger root.

Recommended Dosage and Use: two capsules at mealtime. Female developer. Dong quai has a history of enlarging breasts.

FEMALE PROBLEMS

golden seal, capsicum, false unicorn, ginger, uva ursi, cramp bark, squawvine, blessed thistle, red raspberry.

Recommended Dosage and Use: two capsules at each meal to relieve severe cramps and corrects female problems.

FITNESS IMPROVEMENT

Siberian ginseng, ho shou wu, gentian, licorice, fennel, comfrey, lemongrass, bee pollen, bayberry, myrrh, peppermint, safflower, eucalyptus, capsicum and black walnut.

Recommended Dosage and Use: two capsules at mealtimes as a vitamin and herb supplement to improve physical fitness-especially helpful for athletic type individuals.

FLU AND VOMITING

ginger, capsicum, golden seal, licorice.

Recommended Dosage and Use: eight capsules the first dose and then reduce doses to 4 capsules every hour until a strong ginger taste occurs, then stop. Start with the same dosage the next day if flu exists. For vomiting take 4 capsules at a time until condition stops.

GLANDS

lobelia and mullein

Recommended Dosage and Use: two capsules with each meal to clean out blockages in the gland areas.

GLANDULAR BALANCER

kelp, alfalfa, and dandelion

Recommended Dosage and Use: two capsules with each meal. Glandular balancer thyroid, pituitary, and liver.

HAIR, SKIN, NAILS

dulse, horsetail, sage and rosemary.

Recommended Dosage and Use; three capsules twice daily. Use to promote healthy hair, skin and nails.

HAY FEVER

blessed thistle, black cohosh, scullcap, pleurisy root.

Recommended Dosage and Use: two capsules twice daily, more or less as needed.

HEART

hawthorn berries, capsicum, garlic.

Recommended Dosage and Use: one capsule at morning and evening meal to strengthen the heart and correct heart problems; slowly build up to eight capsules per day. It is also often used to correct circulatory problems.

HYPOGLYCEMIA

licorice, safflower, dandelion, horseradish.

Recommended Dosage and Use: two capsules every four hours to stabilize blood sugar levels.

HYPERGLYCEMIA/PANCREAS

golden seal, juniper berries, uva ursi, mullein, huckleberry, bistort, buchu, comfrey, dandelion, yarrow, garlic, marshmallow, capsicum, licorice.

Recommended Dosage and Use: two capsules at mealtimes to correct blood sugar levels and aid the pancreas to function properly. Is said to have eliminated the need for insulin for many diabetics.

INDIGESTION (See digestive aid)

Food enzyme formula with hydrochloric acid and pepsin.

INFECTION/GLAND

enchinacea, golden seal, yarrow, capsicum.

Recommended Dosage and Use: two capsules three times per day or more as needed to combat a variety of infection problems especially those dealing with glands such as contagious diseases, colds, and the like.

INFECTION FIGHTING

enchinacea, myrrh gum, yarrow, capsicum.

Recommended Dosage and Use: two capsules three times per day or more as needed to combat infection. This herb combination works very similarly to the one mentioned above except this combination is especially formulated for hypoglycemics since it does not contain the herb golden seal.

INFECTION FIGHTING

golden seal, black walnut, marshmallow, lobelia, plantain, and bugleweed.

Recommended Dosage and Use: two capsules with two meals per day (morning and evening) to combat infection. For many people the herb plantain included in this combination is very effective for eliminating almost any kind of infection.

INTESTINAL CLEANSER

comfrey and pepsin.

Recommended Dosage and Use: two capsules before each meal to aid in digestion problems. The combination is said to be able to cleanse the intestinal wall to eliminate deep-seated putrefaction that could be causing other internal problems as well.

IRON (Anemia)

yellow dock, red beet, nettle, burdock, mullein, strawberry leaves, lobelia.

Recommended Dosage and Use: two capsules three times daily.

KIDNEY

uva ursi, parsley, dandelion, juniper berries, camomile.

Recommended Dosage and Use: two capsules as needed to correct a variety of kidney and bladder problems. Helps to dissolve stones.

KIDNEY

golden seal, cedar berries, uva ursi, parsley, ginger, marshmallow, and lobelia.

Recommended Dosage and Use: two capsules with each meal to correct kidney and bladder problems and maintain good health. Also see kidney stones page 102

LIVER

dandelion, red beet, liverwort, parsley, horsetail, birch, lobelia, camomile, blessed thistle, angelica, gentian, goldenrod, yellow root.

Recommended Dosage and Use: one to two capsules three times daily at mealtime to correct liver related problems such as jaundice, age spots, spleen, and so forth. After correcting the problem, reduce dosage to two capsules per day. The liver is an essential blood purifying organ and so a variety of problems are able to be corrected with this useful herb combination.

LIVER/GALLBLADDER/INDIGESTION

barberry, cramp bark, fennel, peppermint, wild yam and catnip.

Recommended Dosage and Use: two capsules before each meal to aid digestion and help the liver and gallbladder function properly. Toxic blood stream, toxic body.

See also page 103 Gall Bladder Fast.

LOWER BOWEL CLEANSER

barberry bark, cascara sagrada, red clover, lobelia, ginger, capsicum, buckthorn, licorice, cough grass.

Recommended Dosage and Use: two to four capsules before retiring, under sever chronic cases two to four capsules before each meal and at night may be needed. Lower Bowel Cleansers are used for a variety of problems to cleanse the body and aid in the detoxifying and healing process.

LOWER BOWEL CLEANSER

cascara sagrada, rhubarb, golden seal, capsicum, ginger, barberry, lobelia, fennel, and red raspberry.

Recommended Dosage and Use: two to four capsules before retiring, under severe chronic cases two to four capsules before each meal and at night may be needed.

LUNG

comfrey, mullein, marshmallow, slippery elm, lobelia.

Recommended Dosage and Use: two capsules with three meals per day to maintain strong healthy lungs. Later reduce dosage to two capsules per day. Since so many different body functions are related to the ability to breathe easily and properly to rejuvenate the blood with necessary oxygen, this is a very helpful herb combination for many people. The combination is used to correct many different lung, breathing and mucous problems.

LUNG/CHEST

comfrey, marshmallow,. lobelia, chickweed, and mullein.

Recommended Dosage and Use: two capsules with all three meals per day again to correct a variety of lung, chest and bronchial problems.

MALE HORMONE

ginseng, damiana, echinacea, gotu kola, saw palmetto, sarsaparilla, periwinkle, chickweed, garlic, capsicum.

Recommended Dosage and Use: one to two capsules at mealtimes to correct male hormone levels. Although it is formulated primarily for males, many females have used it to counteract high hormone levels or to supplement deficient hormone levels in their systems. And it is also used for hot flashes, in women.

MEMORY

gotu kola, capsicum and ginseng.

Recommended Dosage and Use: two capsules with three meals per day to improve memory potential. Used as an aid for many memory problems from such simple things as daily use to increase potential memory in school or at work or for more severe problems such as senility when old age sets in. Also energy and longevity.

MENOPAUSE

black cohosh, licorice, false unicorn, Siberian ginseng, sarsaparilla, squawvine, and blessed thistle.

Recommended Dosage and Use: two capsules with three meals daily to help women through this difficult time when hormone balances are changing so drastically in their system.

MIGRAINE HEADACHES

fenugreek, thyme

Recommended Dosage and Use: three capsules with each meal. Also useful to dissolve mucus in the intestinal tract and cutting fevers.

NAILS (See hair)

NERVES

capsicum, valerian, black cohosh, mistletoe, ginger, hops, wood betony, and St. Johnswort.

Recommended Dosage and Use: two capsules three times per day or less as needed for nerve related problems from nervousness to various types of pain.

NERVES

black cohosh, capsicum, valerian, mistletoe, lady's slipper, lobelia, scullcap, hops, and wood betony.

Recommended Dosage and Use: two capsules with morning and evening meals for nerve related problems. This herb combination seems to work best on the person who is both quiet and hyper at the same time — a person who holds anxieties within.

PAIN RELIEF

valerian, wild lettuce, capsicum.

Recommended Dosage and Use: two capsules with each meal of the day. This is a very simple combination that has proved effective in numerous situations where pain is involved.

PANCREAS/BLOOD SUGAR PROBLEMS

cedar berries, uva ursi, licorice, capsicum, mullein, and golden seal.

Recommended Dosage and Use: two capsules with three meals each day.

PANCREAS REBUILDER

golden seal, juniper, uvra-uris, huckleberry, mullen, comfrey, yarrow, garlic, capsicum, dandelion, marshmallow, buchu, distort, licorice.

Recommended Dosage and Use: 2, three times a day with meals.

Caution: Check blood sugar level regularly while using this formula, as insulin intake may need to be lowered.

PARASITE REMOVER

culver, mandrake, violet, pumpkin, cascara sagrada, witch hazel, mullein, comfrey, and slippery elm.

Recommended Dosage and Use: one to two capsules per meal as needed to remove stomach, intestine, and colon parasites. This combination has a laxative effect and is very effective against parasites in the gastro-intestinal tract. See also Parasite Fast page 102.

PETS, ANIMALS

kelp, alfalfa, dandelion.

Recommended Dosage and Use: approximately two capsules two times per day, depending upon the size and kind of pet. Helps pets maintain the needed vitamin and mineral levels. Is said to be able to keep fleas off most pets.

POULTICE

comfrey, golden seal, slippery elm, aloe vera

Recommended Dosage and Use: this combination is specially formulated to be mixed and used as a poultice to be applied externally to aid healing and soothe painful sores and aches. Use internally as a knitter and healer.

Recommended Dosage and Use: two to three capsules with each meal.

PRENATAL

red raspberry, squawvine, black cohosh, pennyroyal lobelia.

Recommended Dosage and Use: The dosage is usually nine capsules per day in the last fourth and fifth weeks of pregnancy and then to twelve capsules for the last three weeks before birth.

PRENATAL FORMULA

squawvine, blessed thistle, black cohosh, pennyroyal, false unicorn, red raspberry, and lobelia.

Recommended Dosage and Use: three capsules with each meal taken for the last six weeks of the pregnancy.

PROSTATE

kelp, black cohosh, gotu kola, licorice, golden seal, lobelia, ginger, capsicum.

Recommended Dosage and Use: three capsules with each meal to aid with male related prostate problems such as difficulty in urination and the like.

PROSTATE

juniper berries, golden seal, capsicum, parsley, ginger, Siberian ginseng, uva ursi, queen of the meadow, and marshmallow.

Recommended Dosage and Use: three capsules with each meal to alleviate male prostate problems.

REDUCING AID

chickweed, licorice, safflower, gotu kola, mandrake, fennel, echinacea, black walnut, hawthorn, papaya, dandelion.

Recommended Dosage and Use: two to five capsules three times daily before meals to curb excessive appetite and to help relieve the body of acquiring unnecessary weight.

SAFE SLEEP

hops, valerian, scullcap.

Recommended Dosage and Use: two capsules three times per day as a nervine to achieve a relaxed, restful feeling that encourages and aids in sleeping. To promote sleep, four capsules 1/2 hour before retiring.

SINUS

golden seal, capsicum, parsley, desert (Brigham) tea, marshmallow, chaparral, lobelia, and burdock.

Recommended Dosage and Use: two capsules three times per day to aid with sinus related problems such as allergies, hayfever, and sinusitis.

SKIN (See hair)

THROAT, LUNGS, AND HEALING

comfrey and fenugreek

Recommended Dosage and Use: two capsules three times a day to help in a variety of conditions relating to the lungs, throat, and nose primarily. Also very often used with other combinations to achieve special healing results.

THYROID

Irish moss, kelp, parsley, capsicum

Recommended Dosage and Use: two or more capsules as needed for thyroid related problems.

THYROID

parsley, watercress, kelp, Irish moss, sarsaparilla, black walnut, Iceland moss.

Recommended Dosage and Use: two capsules with each meal to maintain proper thyroid functioning.

ULCERS/SORES

golden seal, capsicum, myrrh gum.

Recommended Dosage and Use: two capsules before meals as needed. This combination is used not only as an aid to stomach sores to boils and ulcerations on the outside of the body.

VAGINA

squawvine, chickeweed, slippery elm, comfrey, yellow dock, golden seal root, mullein, marshmallow.

Recommended Dosage and Use: melt cocoa butter in double boiler, add herb powder to make a thick dough-like consistency, cool sufficiently for handling, roll small finger-sized cylinders (on wax paper), let harden and refrigerate, then coat with vegetable oil.

COMBINATION POTASSIUM

kelp, dulse, watercress, wild cabbage, horseradish, and horsetail.

These herbs are high in potassium, other trace minerals and especially silicon.

COMBINATION BON-X

white oak bark, comfrey root, mullein leaves, black walnut leaves, marshmallow, queen of the meadow, wormwood herb, lobelia, scullcap.

The combination of herbs has been used to repair discs, joints, vertebrae, and other degenerative bone conditions such as hip joints and broken bones. Also called the "Bone Knitter Combination."

MYRRH GUM COMBINATION

myrrh gum and slippery elm bark

These herbs have been used to heal mucous membrane tissue such as the alimentary cannel, and the gums in the mouth.

COMBINATION PROTEIN 96

protein, capsicum fruit and red clover tops.

Has been used for a protein supplement between meals for those who need protein frequently as a pick-up.

MILD FOOD DIET

MILD FOOD IN CHRONIC SICKNESS
All Fruits and Vegetables (as raw as possible)
Fruit Juice (canned, raw or frozen)
Vegetable Juice (raw only)
Soft Oil (raw, cold pressed)
All Nuts (must be raw)
Honey (raw)
Sprouts (alfalfa, bean, grains)
All starch vegetables must be baked. Potatoes, squash, parsnips, yams at 500 degrees until done, generally 20-30 minutes.

NO CONCENTRATED FOODS WHEN SICK
Grain
Sugar
Dairy Products
Butter
Eggs
Dried Legumes
Meat
Peanuts
Chips, etc.

The following dietary considerations are listed because we know "you are what you eat!" and "what you eat today, walks and talks tomorrow." If you continue to eat a poor diet and take herbs, do not expect the immediate or desired results that you would like. In acute and chronic diseases we must help the herbs work in the body. It is recommended that the following diet be followed.

For those who wish ultimate health and especially for those in chronic degenerate diseases. No one should live on a diet, but in order to guarantee long and continued health, the following should become your way of life. If you are going to eat the wrong foods, it is recommended that when you do, you do it sparingly and on rare occasions. As it has been said by one great man, Dr. Corwin West, "if you are going to be bad, be real bad and get it out of your system." Then we recommend that you return to your life-giving way of life — with proper foods as listed below:

DIET

CATEGORY	FOODS ALLOWED	FOODS TO AVOID
Beverages	All herb teas (chamomile, comfrey, peppermint, alfalfa, etc.) no caffeine. Fresh fruit juice or frozen fruit juice. fresh vegetable juice	Alcohol, cocoa, coffee, carbonated beverages, canned and pasteurized juices, artificial fruit drinks
Dairy Products	Raw milk, goat milk, yogurt, butter and buttermilk in limited quantities Non-fat cottage cheese and white cheese	All processed and imitation butter, ice cream, toppings. All orange and pasteurized cheeses
Eggs	Poached or boiled eggs (one per day)	Fried eggs
Fish	Fresh white-flesh, broiled or baked	Non-white-flesh, breaded or fried fish or shell fish
Fruit	All dried (unsulfured), stewed, fresh, frozen (unsweetened) fruit	Canned, sulfured or sweetened fruit

Additional Recommendations:

Avoid smoke, exhausts, foods which have been sprayed with pesticides, food additives (especially MSG and others ending in -ate), and foods with artificial colors, flavors, and preservatives.

CATEGORY	FOODS ALLOWED	FOODS TO AVOID
Grains	Whole grain cerals, bread, muffins (e.g. rye, oats, wheat, bran, buckwheat, millet), brown rice, whole seeds (sesame, pumpkin, sun-flower, flaxseed) Nutri-Max pastas	White flour products, hulled grains and seeds, (e.g. pasta, crackers, macaroni, snack foods, white rice, prepared or cold cereals, cooked seeds)
Meats	Only occasionally, then limited amounts of lamb, veal or white portion of chicken or turkey. Meat should be eaten sparingly or only in times of extreme cold or famine	Beef, pork, all prepared meats (e.g. sausage, cold cuts, weiners)
Nuts	All fresh, raw nuts	Roasted and/or salted nuts, especially peanuts
Oils	Cold-processed oils (e.g. safflower, corn) margarine (if safflower or corn oil)	Shortening, refined fats and oils (unsaturated as well as saturated), hydrogenated margarine
Seasonings	Herbs, garlic, onion, chives, parsley, marjoram, capsicum, kelp, vegetable seasoning broth	Black pepper, salt, monosodium glutemate, food enhancers
Soups	All made from scratch (e.g. salt-free vegetable, hicken, barley, millet, brown rice)	Canned and creamed (thickened) soups, commercial boullion, fat stock
Sprouts	All, especially wheat, pea, lentil, alfalfa and mung	None
Sweets	Raw honey, unsulfered molasses, carob, unflavored gelatin, pure maple syrup (in limited amounts)	Refined sugars (white, brown) chocolate, candy, syrups
Vegetables	All raw and not over-cooked fresh or frozen, potatoes parsnips, yams, baked or steamed	All canned vegetables, fried potatoes in any form, corn chips.

HEALING CRISIS AND CLEANSING

HEALING CRISIS

- Happens only as the body is naturally cleansed through fasting and/or semi-fasting and correct body-building foods.
- Happens only when the body has enough vitality to stand the shock.
- Happens when a person feels the best.
- Usually takes about 3 months of correct eating to bring about a healing crisis.
- Only lasts two to three days at most. No need to take enemas or help in any way
- Sometimes, by correct eating, semi-fasting, etc., the body picks off the waste a little at a time and no crisis is necessary.

DISEASE CRISIS

- Happens when the body is too full of mucus and clogged to the limit.
- Happens when enough germs are multiplying.
- Happens when the body strength and vitality are lowest.
- Happens to save the life. If clogging continues at the rate it is going, the person would die because of injury to body organs, poisons in the blood and pumping through the heart, crowding vital organs as in cancer, etc.
- Lasts several weeks.

• Happens sometimes when the body becomes extremely cold, causing the body to squeeze like a sponge, starting an elimination.

• The difference between a Disease Crisis and a *Healing* Crisis is that in a *Disease* Crisis the body only cleanses down to the point where the body can tolerate the poisonous waste and does not completely eliminate it all (this is why diseases reoccur many times); in a *Healing* Crisis the body has the ability to completely eliminate all the toxic waste that the body has at that time.

KIDNEY STONES

Stop all eating.

Juice of one lemon and 1/2 tsp, chlorophyll in four ounces of distilled water every 3 to 4 hours and 4 ounces of lemon juice straight when pain is present.

Golden Seal — approximately 3 to 4 capsules daily, if no low blood sugar. If you have low blood sugar, use Myrrh.

Hot enema when pain is down.

Lobelia or other relaxant tea.

You need to go on a Mild Food Diet during this period of time, may take one day or longer.

PARASITE FAST

Herbal Pumpkin — 3 capsules, 3 times a day.

Black Walnut — 4 capsules, 3 times a day.

Chaparral — 1 capsule, 3 times a day.

Special Formula — 1 capsule, 3 times a day.

For three days eat raw fruits and drink fruit juices (diluted with 1/2 water).

Drink 100% pure fruit juices and if you eat red apples, peel them because they are dipped in wax.

You may take any other herbs or supplements while you are on this, especially Vitamin C.

On the 4th morning, give yourself a garlic enema.

This type of fast works majorly on the intestinal tract only. Many people work many months eliminating parasites from more difficult parts of the body.

GALL BLADDER FAST

• Eat a light breakfast such as two slices of whole wheat toast.

• Prepare a drink by squeezing 1/2 of a lemon into 4-6 ounces of distilled water and drink at least every two hours. Continue fasting and just drinking this drink throughout the rest of the day.

• At approximately supper time consume a strong dose of an herbal laxative such as Cascara Sagrada, continuing drinking your lemon juice and distilled water.

• At bedtime prepare 4 ounces of fresh squeezed undiluted lemon juice in one cup and in a separate cup pour 4 ounces of unheated pure olive oil.

NOTE: Do not mix together as it is nearly impossible for a person to drink 8 ounces of this solution all at one time without gagging or possibly vomiting it back up.

• Take these two cups to your bed. While sitting on the side of the bed drink the lemon juice straight down, lie down immediately on your right side for a minimum of 15 minutes.

• Sit back up and at this time drink down the 4 ounces of oil. Some people save maybe a teaspoon or tablespoon of the lemon juice to drink at this time to cut the oily taste in the mouth. Immediately after drinking the oil, however, lie down on the right hand side for a minimum of two hours. You can even put a pillow under the hip if you desire.

Note: If you do not follow these instructions and you get up before the two hours, you may have an emetic situation and then you add insult to injury because you have to start all over again. You should allow the solution to circulate in your body at least 8 hours before the next step. Most people go on to sleep.

• Upon arising the next morning, drink 1/2 cup of warm water with 1/8 teaspoon of salt in it and go take an enema. You can lay a towel or old blanket on the bathroom floor to make yourself comfortable and to absorb any accidents.

The water should be at least 5 degrees warmer than body temperature; the bag should be no higher than one foot above the buttocks' area. Water should be taken very slowly. Relieve yourself when there is discomfort.

When having an evacuation, stones will be found floating: may be a very darkish color until washed off and they generally are a greenish color of various sizes; may be scooped out with a strainer. If saving, put in a bottle with water with a lid on them.

Enema should continue until the individual has taken the full two quarts all at one time and water comes out clear.

• The gall bladder fast is recommended to be done at least three times or until no stones come forth. Stones may even be found during normal eliminations in the latter part of the day.

Fasting isn't uncomfortable neither is the enema, both are relaxing and both are of great benefit to the body in purifying it internally.

IDEAL SOLUTION FOR A HIGH ENEMA

Take 3 buds or 2 to 3 capsules of garlic; put in a blender with three cups of distilled water; blend until liquid. Use 1/2 cup of solution to 2 quarts of warmed water.

The addition of 1/2 teaspoon to one teaspoon of an organic soap which is of a surfaction can greatly help in softening. Sometimes the addition of an herb such as Lobelia or Chamomile is also very relaxing.

Do not worry if the water does not come out immediately or at all many times: the retaining of water is helping with dehydration, although you can use the last water entered into the body at a cooler temperature, which will cause rapid expulsion, if need be. Relieve yourself when there is discomfort.

HIGH ENEMA (HIGH COLON IRRIGATION)

There are many positions you can take your enema in. One is called the knee-to-chest position with buttocks in the air.

The other is you start the enema by lying on the left side first until water can be felt up over the left rib cage; then roll over on back until water can be felt up over the right corner of the rib cage; and then lie on the right hand side feeling the water going down the right hand side, and then you have completed a high colon irrigation.

Again, relieve yourself when you feel discomfort. There is no way you can take large amounts of water until waste material has been removed, making room for the water.

Note: Cold water causes contractions and hot water relaxes, allowing more retention for a longer period of time, creating a soaking and loosening action. The water should be 5 degrees above body temperature and the bag should be no higher than one foot above the buttocks' area.

HERBAL PREPARATION

Herbalists have always been concerned about people not taking enough of an herb to get the job done. The standard nutritional guideline of the need of five to seven times the normal amount of nutrition to rebuild and repair has always been difficult for the lay person to comprehend. The fact that they were asked to take large amounts of what they considered to be medicine always worried them. If we are talking about the standard allopathic world, there is certainly cause for concern. We hear almost every day about someone overdosing on pharmaceutical drugs whether they be over-the-counter or prescription. Generally the herbalist's problem has always been in getting the person to take enough and long enough to get the job done. It is recognized that there is little if anything of any importance that will heal itself in the human body in less than three months. I emphasize "heal," bringing the body to a normal healthy state, not to just simply supress the pain or misery that a person is going through. It is this author's opinion that one of the greatest boons to mankind, as far as herbs are concerned, has been the encapsulated herb, not only for the convenience of no mess, but no bad taste also.

There are many ways that we can prepare herbs. We can make them into decoctions or infusions or we can simply just say "teas." This method has been used for many hundreds of years and is still the most common way of taking herbs into the body. This method does have its draw-backs though, such as taste. Many of the medicinal herbs come from a group of herbs called the "bitters," and they earn their title quite honestly. Second problem here is that you are applying high heat to a live substance and at 120 degrees even the minerals are going. And we can only extract the water soluble vitamins and minerals with this preparation. So we are not getting a complete, whole food.

We must stress at this point the effectiveness and ease of assimulation because the herb is now in a liquid form.

There certainly are tinctures to consider. Their shelf life is indefinite, and their absorption quality is high whether used topically or ingested. The alcohol can even be assimilated through the walls of the stomach as very few things can. Again we have the draw-back of the bad taste and those who have an adversion to alcohol.

We now seem to see the light at the end of the tunnel and seem to be led to the end of that tunnel by the Chinese. Unlike the Western world, Chinese did not discard herbal medicine for drugs. At least 80% of the medicines used in China are considered to be natural remedies. I think it is quite appropriate that the Chinese have led us to a great improvement in the method of taking our herbal medicine.

Here in the U.S. some herb capsule manufacturers have taken a leaf from the Chinese and are turning to a new process that makes very concentrated herbal extracts for filling their capsules. In this way it is easy to get the full potency of what one needs without having to take a large handful of capsules many times a day. To make their new concentrates, the herbs are first extracted and alcohol is generally used for this process, after filtering the cellulose starch and other inert ingredients the extract is dried to form a very concentrated powder. A single gelatin capsule can hold the dried extract from several grams of herbs. In this way a small number of capsules can provide the full potency of the single herb or combination of herbs that is needed to get the job done.

The major difference is that they do cost more, but because a single capsule of the concentrate does contain generally all that is needed for an individual during one of the periods that they will be taking herbs in a day, it is not necessarily more expensive. Whether for nutritional, tonic, or curative purposes, the potency of an herbal treatment is a very important consideration. The extra cost of concentrates is certainly a worthwhile investment.

BIBLIOGRAPHY

Aikman, Lonnelle, 1977. *Nature's Healing Arts: From Folk Medicine to Modern Drugs*, 200 pp. National Geographic Society.

Chai, Mary Ann P *Herb Walk Medicinal Guide*. Provo, UT: The Gluten Co., 1978.

Coon, Nelson. *Using Plants for Healing*. Emmaus, PA: Rodale Press, 1979.

Farwell, Edith Foster. *A Book of Herbs*. The White Pine Press. Piermont, NY, 1979.

Goulart, Frances Sheridan. *The Grey Gourmet*. Thornwood Books, 1980.

Heinerman, John. *Science of Herbal Medicine*. Orem, UT, Bi-World Publishers, 1979.

Hylton, William H., ed. *The Rodale Herb Book*. Emmaus, PA: Rodale Press, 1975.

Kloss, Jethro. *Back to Eden*. Santa Barbara, CA: Lifeline Books, 1975.

Krochmal, Arnold and Connie. *A Guide to the Medicinal Plants of the United States*. Quadrangle Books. NY, 1973.

Krochmal, Connie, 1980. *A Guide to Natural Cosmetics*. 225 pp. Thornwood Books, Springville, UT.

Millet, Edward Milo. *Herbal Aid*. Thornwood Books, 1980.

————. *Herbs for Building Health and Vitality* (Vols. I, II, III). Thornwood Books, 1980.

Moulton, LeArta, 1979. *Herb Walk 1*. The Gluten Co., Provo, UT.

Rose, Jeanne, 1972. *Herbs & Things*. Grosset & Dunlap Workman Publishing Company, NY.

Royal, Penny C., 1976. *herbally Yours*. 128 pp. Bi-World Publishers, Inc., Provo, UT.

The Little Vitamin and Mineral Encyclopedia. Thornwood Books, 1980.

To Your Health: The Best of the Herbalist Magazine 1975-1980. Thornwood Books, 1980.

INDEX

NOTE: Herbs are listed in capital letters. Bold type indicates major treatment of subject.

A

B

C

D

E

F

G

H

I

J

K

L

M

N

O

P

Q

R

S

T

U

V

W

Z